Femtocells: Secure Communication and Networking

RIVER PUBLISHERS SERIES IN COMMUNICATIONS

Consulting Series Editors

MARINA RUGGIERI
University of Roma "Tor Vergata"
Italy

HOMAYOUN NIKOOKAR
Delft University of Technology
The Netherlands

This series focuses on communications science and technology. This includes the theory and use of systems involving all terminals, computers, and information processors; wired and wireless networks; and network layouts, procontentsols, architectures, and implementations.

Furthermore, developments toward new market demands in systems, products, and technologies such as personal communications services, multimedia systems, enterprise networks, and optical communications systems.

- Wireless Communications
- Networks
- Security
- Antennas & Propagation
- Microwaves
- Software Defined Radio

For a list of other books in this series, visit
http://riverpublishers.com/river_publisher/series.php?msg=Communications

Femtocells: Secure Communication and Networking

Marcus Wong
Huawei Technologies, USA

River Publishers

Aalborg

ISBN 978-87-92982-85-8 (hardback)

Published, sold and distributed by:
River Publishers
P.O. Box 1657
Algade 42
9000 Aalborg
Denmark

Tel.: +45369953197
www.riverpublishers.com

Contents

Preface   **ix**

I   Getting Started   **1**

1   Introduction   **3**
1.1 1G and Analog Systems 5
1.2 2G Systems 7
    1.2.1 TDMA Systems 8
    1.2.2 CDMA Systems 9
    1.2.3 Other 2G Systems 10
1.3 3G 11
    1.3.1 UMTS 12
    1.3.2 CDMA2000 12
    1.3.3 TD-SCDMA 13
1.4 4G 13
    1.4.1 HSPA 14
    1.4.2 LTE 15
    1.4.3 WiMAX 16
1.5 Femtocells 16

2   3GPP Architecture and the Femtocell   **19**
2.1 UMTS Architecture 19
2.2 LTE 22
2.3 Femtocells and Femtocell Architecture 26

II   Securing the Femtocells   **31**

3   Security of Femtocells   **33**
3.1 UMTS and LTE Security 34
3.2 Femtocell Security 36

3.2.1 Femtocell Threats and Attack Model...............................36
3.2.2 Requirements 57
3.2.3 Femtocell Security Mechanisms 66
3.2.4 Security of UE Mobility in Femtocells...........................84
3.2.5 Femto Security and LIPA 87
3.2.6 Emergency Services Support in Femtocells.....................88
3.2.7 Femtocell Security Profiles...88
3.3 Case Study on Attacks of the Femtocells...............................88
3.3.1 Case 1...89
3.3.2 Case 2...89
3.3.3 Case 3...89
3.3.4 Case 4...90
3.3.5 Case 5...91

4 CDMA Femtocells   **93**
4.1 Variations in CDMA Femto Architecture 98
4.2 CDMA Femtocell Security..100
4.2.1 FAP Security Features and Mechanisms 101
4.2.2 Authentication..104
4.2.3 Authorization 105
4.2.4 Integrity and Confidentiality Protection 106
4.2.5 FMS Security 107
4.3 Security Mechanisms and Procedures.....................................109
4.4 Differences between CDMA and UMTS/LTE Femtocells 111
4.4.1 SIM Card Support..111
4.4.2 Use of Pre-shared Key 112
4.4.3 Use of Device Certificate...113
4.4.4 FAP Authorization 113
4.4.5 Signed File Transfer...113
4.4.6 Optional Use of IPsec 114
4.4.7 Direct Interface between FAPs 114
4.4.8 CSG Handling...115
4.4.9 FAP Identity Verification 115
4.4.10 UE or MS Authentication 116
4.4.11 LIPA Access 117
4.4.12 Security Profiles...117
4.4.13 Other Differences...117

5 WiMAX Femtocells   **119**
5.1 WiMAX Architecture and the Femto.....................................119

5.2 WiMAX Femto Functional Components 121
    5.2.1 Femto Access Points (WFAP) 121
    5.2.2 Security Gateway (Se-GW) 123
    5.2.3 SON 124
    5.2.4 WFAP Management System 126
    5.2.5 Bootstrap server 126
    5.2.6 Femto-AAA Server 127
    5.2.7 Femto Architecture Reference Points 127
5.3 WiMAX Femto Security Features and Mechanisms 129
    5.3.1 WiMAX Femto Access Point Initialization 129
    5.3.2 WFAP Network Exit 133
    5.3.3 Closed Subscriber Group Management 134
    5.3.4 Mobility Management 135
    5.3.5 Network Management Procedure 135
5.4 Differences between WiMAX and 3GPP/CDMA Femtocells . 136
    5.4.1 SIM Card Support 136
    5.4.2 Use of Pre-shared Key 136
    5.4.3 Use of Device Certificate 136
    5.4.4 Optional Use of IPsec 137
    5.4.5 Direct Interface between FAPs 137
    5.4.6 CSG Handling 137
    5.4.7 FAP Identity Verification 137
    5.4.8 LIPA Access 138
    5.4.9 Security Profiles 138
    5.4.10 Other Differences 138

6 LIPA and SIPTO   **139**
6.1 Security Considerations in LIPA and SIPTO 141
    6.1.1 Legal Intercept 141
    6.1.2 Billing and Charging 143
    6.1.3 Backhaul Security 144
    6.1.4 Mobility Support 145
    6.1.5 L-GW to Femto Interface 145
6.2 In Short 146

7 Security Profiles and IKEv2 Call Flow   **147**
7.1 Security Profiles 147
    7.1.1 TLS Certificate Profile 147
    7.1.2 IKEv2 Certificate Profile 149
    7.1.3 TR-069 Protocol Profile 150

7.1.4 IKEv2 Usage Profile 151
7.1.5 TLS Usage Profile 152
7.2 IKEv2 Example Call Flow Used in 3GPP 152

III From Femtocells to Small Cells   **161**

8   From Femtocells to Small Cells   **163**
8.1 Small Cells 163
8.2 Small Cells and Wi-Fi 167

IV Outlook and Concluding Remarks   **171**

9   Conclusion and Outlook   **173**

Annex   **177**
A.1 Sample of TR-069 FMS Security Related Parameters 177
A.2 WiMAX SON 200
A.3 CDMA2000 Authentication 202
    A.3.1 Global Challenge 203
    A.3.2 Unique Challenge 203

Glossary   **205**

Bibliography   **213**

About the Author   **217**

Preface

What is a Femtocell and how secure is it? This book contains detailed answers to both parts of this question.

This book deals with past, present and the future of the wireless communication system and the Femtocell. An introduction chapter will take the readers through a journey of wireless communication from the past to the present and from the analog systems that was referred to as first generation or 1G all the way to the latest and greatest in fourth generation or 4G. Chapter 2 provides an architecture overview of the UMTS and LTE systems and the Femtocell that is based on the two technologies. Chapter 3 through 7 provides in depth coverage of the security of Femtocells based on various access technologies, including UMTS/LTE, CDMA, and WiMAX. Security of other aspects related to the Femtocells such as LIPA, SIPTO, and emergency call will also be discussed. Chapter 8 deals with small cells that have become so popular with the advances in Femtocells. And finally, the author gazes through the crystal ball and put forth some concluding remarks on the future prospect of Femtocells and small cells in general in Chapter 9.

The book is an outgrowth of the author's experience in the telecommunication and security field that had started in the early 1990s. Contents of the book are based on knowledge and insight gained by working with industry security experts, customers, and many other engineers in the field who have deep knowledge and understanding of wireless communication systems as well as research and contribution to the standardization of Femtocell security in 3GPP. The book has been written to provide working knowledge and insight in the security design of Femtocells for mobile network and device architects, designers, researchers, and students alike.

As the Femtocells and small cells are still evolving and are subject to constant changes, readers are encouraged to provide comments, questions, and suggestions for future edition.

The author thanks Dr. Anand Prasad, without whose introduction of the editors from River Publishing would have made it impossible to publish this book. His many suggestions and comments are invaluable. A special note of thanks goes to Rajeev Prasad of River Publishing for giving the author the opportunity and a medium to share the research and knowledge that went into designing the security of the Femtocells.

Part I

Getting Started

1

Introduction

Telecommunication systems have come a long way ever since the days of using fire and smoke as a way to communicate and signal across long distances. In ancient China, the soldiers stationed along the Great Wall of China's towers would use this mechanism to signal the coming of the invaders along the way. Soldier in the first tower that observed incoming invaders would build the fire that generated large amount of smoke that could rise high above the sky so that other soldiers stationed in adjacent towers along the Wall would be able to see and they themselves would in turn create another fire with the same effect so that other soldiers would be warned. It is said that with this system of warning, a message could be sent as far as a few hundred miles in the matters of hours. However, this primitive communication system has many drawbacks, with the biggest being the limit to which a message can be relayed. This is the equivalent of communicating just one-bit of information of a modern day communication system. Another drawback is the distance and speed of which the smoke signals can travel as the naked eye can only see so far. Many relays have to be set up along the way of the communication path. Another improvement is made with the use of pigeons that can carry longer messages than simply yes or no. But that too had many drawbacks as well. It took a couple thousand years, but advances in technology have just about made both the fire and smoke as a communication system and the pigeon system obsolete. The invention and improvement of the telegraph system in the 1700s and 1800s have greatly improved the speed and distances in which people communicated. Over the next 150 to 200 years, even greater advanced made it possible for people to communicate without a wire. It was during a time of great scientific advancement that the cellular system was born as an invention out of the old Bell Laboratories in the 1950s. The cellular system uses spreading relay stations positioned strategically similar to that of a beehive and connected to the network that can be used to relay

3

telephone signals through the network. These so-called relay stations become the base stations and they can communicate wirelessly with a mobile unit. In turn two mobile units communicate to each other through the base station. In order to provide coverage over a large area, the base stations are strategically located in a cellular fashion that resembled a beehive, which provides overlapping communication coverage for the mobile units on the ground. This beehive-like cellular concept had its roots back in the late 1940s when the first small base station began connecting to mobile phones. As the cellular concept matured and flourished under constant research and improvements, re-use of frequencies became feasible and later became a must when large area is to be covered with base stations with relatively small amount of spectrum. Since the 1980s, cellular technology has made its way to be the communication system of choice for people on the go, making communication with anyone from anywhere at any time possible.

To communicate, mobile unit and the base station send and receive signals over the air using spectrum of frequency. Different frequencies are used for sending and receiving, thus avoiding signal interference. Since the spectrum frequencies are limited and at a great premium, they are tightly controlled by regulatory agencies. Careful planning and deployment are essential to maximize the spectrum efficiency which caused the network operators to deploy initial networks to densely populated areas initials. As more and more mobile users are put into the network, the network operators increasingly find that they are running out of spectrum. Though many technologies and effective use of spectrum have been identified and put into use, the operators are still facing dilemmas of extending coverage, providing new services, and at the same time lowering overall cost of network deployment.

There is a finite amount of spectrum available to the operators and the spectrum is acquired by operators paying billions of dollars through legal agreements (i.e. spectrum licenses) with the regulatory agency. Each operator that purchased the spectrum must obey by regulation that it only can operate the spectrum in the geographic areas or markets the spectrum is designated for. If the operator operates the spectrum where the license is not valid, the operator is subject to fines including and up to revocation of the license that it is granted. Because of this, wireless networks tend maximize the use of spectrum by providing the best coverage in densely populated areas where most of the users are. The few users who are not well within the covered area or are on the edge of the network often experience poor signaling and call quality due to the lack of coverage. Even in areas of good coverage, a user may still experience pool call quality due to other reasons such as network congestion, signaling interference from other users, other devices, or topological differences in deployment environments. Worse, a network may simply not able to handle the number of users that far exceeds the volume that the network was originally designed to carry. In order to maintain and expand the user base and the same time to keep the user satisfaction at a high level, the operators and network equipment manufacturers are forced to come up with ingenious ways of extending the operator network coverage at a reasonable price—the Femtocell is born.

Let us take a stroll through the history of how cellular system has evolved from the beginning with the analogue system first before returning to the Femtocell.

1.1 1G and Analog Systems

There had been many attempts to marry the wireless communication with the telephone ever since the invention of the telephone by Alexander Graham Bell in 1876. The first marriage of wireless and the telephone was reported as early as in the late 1918 when the Germans experimented with the mobile telephone on military trains between Berlin and Zossen. Various trials and improvements were made between then and 1946 by scientists and engineers in both real world applications as well as in the laboratories. Though the first commercial mobile telephony service was offered as early 1946, but those early systems were not necessarily classified as cellular systems as the frequencies and technology to make it a true cellular system were not available at the time. The advances in cellular concepts were being worked out at the AT&T Bell Laboratories soon after the first commercial mobile telephone service. The idea is divide the wireless coverage into hexagonal cells:

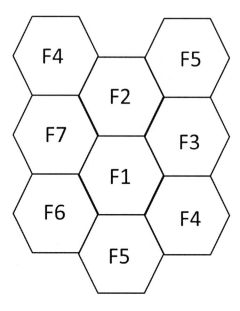

Figure 1.2. Cellular Concept with 7-cell Reuse Factor

Each cell would use a different frequency to send and to receive than the frequencies that are used the adjacent cells. When the distance is far enough so that the frequency interference is not an issue, the frequencies would repeat in another cell. Figure 1.1 shows a 7-cell cellular concept where the frequency re-use factor is seven. The cellular concept greatly improved the reuse of limited spectrum resource, which was the limited factor to offer wide range mobile telephony service at the time and still remain so at present day.

The first mobile telephones, including and up to what was known as 1G cellular service, used analog technology. The analog technology uses separate frequencies, or channels, for each conversation between the cell tower (e.g. base station) and the mobile phone. Because of the amount of frequency available for each cellular system, the use of different frequencies to carry different conversations meant that the system was not able to support many simultaneous users, even with the cellular concept of re-using frequencies at different cells. In addition to the limited capacity of the system, there were other inherent issues with analog transmissions, among them being static, noise, eavesdropping via scanning, etc. that all contributed to the shortcomings of the 1G systems.

As the number of independently operated mobile telephone system proliferated following the commercial launch, the industry got together and started to work on bringing the players together to come up with standards for better interoperation and user experience. Though the initial motives for the standards were less noble and the process was both political and technical in nature, but it did bring the entire industry together. In the US, Advanced Mobile Phone System or AMPS became the 1G standard in the early 1980s. The AMPS standards were also adopted in many parts of the world, such as Australia, Israel, and rest of the Americas. At the same time, the standards process was so disjoint that different regions and/or countries also got together and developed their own standards. There were several other such 1G standards that existed. Among those are:

- Nordic Mobile Telephone (NMT) used in Nordic Countries, Eastern Europe and Russia
- Total Access Communication System (TACS) used in United Kingdom, West Germany, Portugal and South Africa
- Radiocom 2000 used in France
- Radio Telefono Mobile Integrato (RTMI) used in Italy
- TZ-801, TZ-802, TZ-803, and JTACS used in Japan

Essentially, the basic characteristics of the first generation analog cellular systems were mainly low speed download, ranging from 28 kilobits/second to 56 kilobits/second, low network capacity, and low security. Additionally, high cost of both network equipment and handsets were a deciding factor for its limited popularity as yet the industry saw the potentials. They plowed on.

1.2 2G Systems

As more and more users are warming to the idea of mobile telephone both as a luxury and as a convenience, the capacity of the first generation systems soon reached its peak. Engineers at the same were busy researching and developing the second generation of the mobile telephone systems. In one direction, they found out that if each channel was divided into multiple time slots, the system capacity can be greatly expanded, with factor of up to eight times the first generation analog systems. The division of multiple time slots worked like this. For a division of two, the sender sends transmits two messages at two distinct times, for example, once every half second to receiver A and once every second to receiver B. The receiver A then tunes itself to receive the message every half second and receiver B tunes itself to receive the message every second. Since receiver A knows that the messages sent on every half second belongs to him, he only needs to capture the message that was sent on every half second and the same applies to receiver B for the message sent on every second. This concept is known as Time Division Multiple Access or TDMA.

In another direction, the engineers found out that supposed a message is encoded with a code for a particular user, and with the proper code filter, only the user who possesses the code can recover the message. This way, the transmitter can use many codes and simultaneously transmit multiple messages so long as each message transmitted is encoded with a unique code only the receiver with that particular code can recover the message. This new concept is known as Code Division Multiple Access or CDMA.

Both TDMA and CDMA technologies are considers as spread-spectrum access technology where the amount of information being sent over a particular frequency is much larger than that of the bandwidth of the frequency itself used by the system. Cellular systems based on the TDMA and CDMA access technologies served as bases for the second generation wireless systems all over the world as technology based on analog took a backseat on its way to distinction. In reality, it took more than a decade before many operators finally pulled the plug on the old analog systems when it became no longer feasible to support both the analog system and a digital system at the same time.

1.2.1 TDMA Systems

1.2.1.1 TDMA

IS-54 and IS-136 are standards developed by the Telecommunication Industry of America (TIA) in North America and subsequently used in both North America and South America based on TDMA technology. These systems are considered evolution of the AMPS system where similar channel structures and spacing were used between the channels.

In TDMA, each channel is divided into three time divisions or slots. Digital capabilities and advanced made in the 1980s and early 1990s made TDMA possible. The subdivision of each channel improved the capacity of each cell, now can serve as many as three times more users than before.

1.2.1.2 GSM

The Global System for Mobile Communication or GSM was the second generation mobile system developed in Europe that united most of Europe as the replacement of varying first generation mobile systems. The system gained such popularity around the world that eventually got adopted in many more parts around the world and became the dominant 2G systems of choice. The GSM systems were TDMA based systems. Since the designer did not have the constraint of being backward compatible with the many 1G systems at the time, the design started from a clean slate. The technology is similar to the TDMA (e.g. IS-54), but used a much wider bandwidth for its channels. Instead of the 30 KHz channels used in the America TDMA systems, GSM used 200 KHz channels. Using the wider channel gave the GSM an opportunity to afford eight time slots for each channel. In terms of security, the first use of a Subscriber Identity Module (SIM) was introduced, along with the security features of GSM that offered users authentication and privacy. The SIM housed the subscriber credentials that are used for authentication and subsequently privacy protection. The SIM concept was so popular that it extended beyond GSM well into 3G and 4G systems.

1.2.2 CDMA Systems

While TDMA enjoyed initial success in the America's, the competing CDMA technology also enjoyed its success in America. During its infancy, CDMA seemed to be more promising, with claims that touted potential capacity improvement as large as twenty times that of the AMPS systems. But that soon proved to be difficult to achieve outside of the laboratories where interference and noise can be isolated in a perfect environment. However, more realistic gains were on the order of three to seven when different factors, included interferences, were accounted in real world environments. Though, in theory, was still better than that of the TDMA, but the technology still needed much tweaking and constant adjustment at the time. To allow the technology to grow, the IS-95 standards published by the same TIA competed with the TDMA standards IS-54 and IS-136 in the US and in Canada (and later in other parts of the world). With advances in both digital signal processing and improvement in coding theory, CDMA soon proved to be more superior to the TDMA counterpart. In addition to offer better capacity gains eventually, it also offered better power usage in terms of longer battery life and therefore longer conversation for the users, and more importantly at the same time, it also offered inherent better security in terms of user privacy. The improvement in security was attributed to the use of cryptographic enhancements in the 2G standards as well as to the inherent security features in CDMA.

1.2.3 Other 2G Systems

At the same time TDMA, CDMA, and GSM gained popularity as the system of choice for most, there were also several other 2G systems used by various operators. One of such system was the Personal Digital Cellular or PDC. It was developed and used exclusively in Japan that used both a three- and six-time slots for both full-rate and half-rate transmissions respectively. Another such system was the Integrated Digital Enhanced Network or iDEN. iDEN was also a TDMA-based technology that also offered walkie-talkie-like features that gained popularity in the US and in parts of South America for many delivery drivers, limo drivers, and the like. The walkie-talkie features that gave users instant access to other users made short calls simple and quick. Instead of dialing a number, the users can simply push a button to communicate with another user instantly. Many iDEN networks have since ceased operations in many parts of the world, including such a system that had successfully operated for many years in the US and was later acquired by one of the competing network operators. But the walkie-talkie features were liked by many users for their convenience and later became a push-to-talk service (e.g. Qchat and vchat) that are offered in many 3G systems and beyond.

1.3 3G

While 2G greatly improved upon capacity, reliability and security of cellular systems compared to 1G, it still cannot catch up with the explosive growth of the number of subscribers. As the number of services that were available in 2G grew as well, issues with capacity and bandwidth were soon found to have become the bottleneck once again. In the late 1990s, the International Telecommunication Union (ITU) set out to define the next generation of cellular communication systems known as the IMT-2000, or International Mobile Telecommunication 2000. IMT-2000 (and later IMT-Advanced) is defined by ITU that has been instrumental in coordinating the efforts of government and industry and private sector in the development of a global broadband multimedia international mobile telecommunication system. Since 2000, the world has seen the introduction of the first family of standards derived from the IMT concept. These families of 3G systems with different access technology are intended to provide a global mobility with wide range of services including telephony, paging, messaging, Internet and broadband data with data speed ranging from 200 Kbits/second to 14.4 Mbits/second. The process started in 1998 when the Third Generation Partnership Project (3GPP) was born and began to work on the new 3G standards based on the evolution of GSM. Around the same time, another partnership – Third Generation Partnership Project 2 (3GPP2) was also born to mirror the 3GPP counterpart and was to set out to develop yet another 3G standards that was based on the CDMA technology. The standards that were published by 3GPP and 3GPP2 were known as UMTS and CDMA2000 respectively and became the predominant 3G standards in many parts of the world. However, IMT-2000s definition was broad and flexible enough that it allowed other competing access technology to be called 3G as well, namely, TD-SCDMA was allowed to be branded 3G in China, DECT systems used primarily in parts of US, Canada, and Europe, and WiMAX, some of which were politically driven as opposed to market driven. Speaking of market driving forces, TDMA-based technologies had fallen out of favor of the operators and manufacturers since 3G evolution began and as the operators and vendors aim to converge in terms of common technology for better interoperability.

1.3.1 UMTS

UMTS or Universal Mobile Telecommunications System was a CDMA-based system. By the time ITU defined IMT-2000, the inherent advantages of CDMA were readily seen by many as the foundation for the next generation of wireless cellular communication services (another reason why TDMA was falling out of favor faster than expected). Although UMTS differed from GSM greatly in terms of air interface, but it was still considered an evolution of GSM, as many features of GSM and interim improvements based on GSM (GPRS, EDGE, etc.) were retained, refined, and improved. The five main areas of UMTS standardization were Radio Access Network, Core Network, Terminals, Services and System Aspects, and GERAN (or GMS EDGE Radio Access Network). The first release of the UMTS specification was published in 1999 and the first UMTS system was deployed just as quickly in 2001.

1.3.2 CDMA2000

CDMA2000 is a family of 3G standards that was truly evolution of the cdmaOne, or IS-95 based systems that provided total and complete backward compatibility. Designed in incremental fashion, CDMA2000 started as CDMA 1xRTT (one time Radio Transmission Technology with one single 1.25MHz bandwidth for transmission and receiving) and can be increased to support up to three carriers (i.e. 3x RTT) if an operator has enough frequency allocation in 5 MHz bandwidth chunks instead of the usual 1.25 MHz chunk of bandwidth. The efficiency of the CDMA2000 air interface designed, lack of contiguous bandwidth allocation, the expense of 3G spectrum license almost made sure that the use of three carriers is unnecessary and most operators adopted to use a single carrier for their CDMA2000-based 3G network deployments. CDMA2000 also supported a data only mode in a standard called the CDMA 1xEV-DO (Single Carrier Evolution Data Only) and another data and voice standard called the CDMA 1xEV-DV. 1xEV-DV was meant to replace 1xRTT and 1xEV-DO, but market conditions and the decision of the 3GPP2 to move toward a common technology as the one chosen by 3GPP for the next generation of mobile communication systems meant that 1xEV-DV was not favored by the operators, let alone having been deployed. For faster data access, many operators deployed

1xEV-DO network that saw data speed of up to 14.4 Mbits/second, which was on par with that of the UMTS networks.

1.3.3 TD-SCDMA

TD-SCDMA or Time Division – Synchronous Code Division Multiple Access was yet another 3G standard that was developed in China by the Chinese Academy of Telecommunications Technology (CATT), Datang Telecom, and Siemens as a 3G technology in an attempt to have a technology that can be used in China without the dependency to western technology such as UMTS or CDMA2000. This technology used both a combination of TDMA and CDMA access technology—it uses synchronous CDMA channel access method over multiple time slots. In China, the market climate and the late granting of 3G spectrum license made the initial deployment on a wide scale after the operator that obtained the 3G license based on TD-SCDMA technology in 2009. As of late, there is only one TD-SCDMA operator in China and the estimated number of users that have been migrated to this faster 3G network is around 140 million out of 700 million in its network.

1.4 4G

Almost as soon as 3G started rolling out, the researchers and engineers had started to work on 4G. It was easy to envision that as 3G deployments and usage were reaching the peak, the operators and users were increasing finding it difficult to satisfy the demands of faster and faster access and fatter and fatter bandwidth. Ubiquity of Internet access and wireless communication has made the natural connection of mobile broadband. People are hungry for information, and specifically instantaneous information. People are hungry for sharing: sharing of data, photos, videos, texts, etc. People are hungry for speed and the same Internet experience on the mobile phones as if they were home. It was the same insatiable appetite of users' desire that made 4G both a possibility and a reality. ITU had set out and defined 4G as the having the following minimum requirements:

- It must be based on an all-IP packet switched network
- It must support peak data rates of up to 100 Mbit/second for high mobility such as mobile access and up to approximately

1 Gbit/second for low mobility such as nomadic/local wireless access.

- It must be able to dynamically share and use the network resources to support more simultaneous users per cell
- It uses scalable channel bandwidths of 5–20 MHz, optionally up to 40 MHz
- It must have peak downlink spectral efficiency of 15 bit/s/Hz and uplink spectral efficiency of 6.75 bit/s/Hz
- For indoor systems, it must have downlink spectral efficiency of up to 3 bit/s/Hz/cell and uplink spectral efficiency of up to 2.25 bit/s/Hz/cell
- It must support smooth handovers across heterogeneous networks, for example from UMTS to CDMA systems
- It must have the ability to offer high quality of service for next generation multimedia support

Currently, there are two families of standards that fit the bill: LTE and WiMAX. Though HSPA and HSPA+ are marketed by many operators as 4G services in many parts of the world, but they are not considered as 4G in the true sense.

1.4.1 HSPA

HSPA or High Speed Packet Access, along with HSPA+, is an enhancement of UMTS that offered improved data rate over UMTS, with uplink speed of up to 22 Mbits/second and downlink speed of up to 168 Mbits/second. Among some of the other noticeable improvements were improvements made on the protocols used both the uplink and downlink, referred to as HSDPA and HSUPA respectively. When WiMAX was first offered as a 4G (see below) service as early as in 2008, many traditional mobile operators around the world that had already deployed UMTS networks or CDMA2000 networks felt that they are threatened by the fact that someone else launched 4G first and that they were lagging behind in their 4G service offering potentially made their service as being inferior to that of WiMAX began upgrading their 3G networks to support HSPA and subsequently marketed such as 4G service. Though technically satisfying many of the requirements set out by ITU, HSPA and HSPA+ were also collectively viewed as 3.5G and 3.75G respectively by many researchers and engineers who were deeply involved in the design and improvements of the existing 3G systems on their way to true 4G.

1.4.2 LTE

LTE/SAE or Long Term Evolution/System Architecture Evolution was truly considered the 4G migration of UMTS-based 3G, CDMA2000-based 3G as well as TD-SCDMA-based 3G. It was truly an amazing international phenomenon that all three of these standards were merged into one in an unity of tour de force when operators that had deployed CDMA2000 and TD-SCDMA also made pledges to migrate from 3G to 4G LTE. In all started in November of 2004, when 3GPP began a project to define the long-term evolution of UMTS cellular technology with increased performance, backward compatibility and support for wide range of new and anticipated applications.

In terms of data rate support, the goal of the LTE was to have peak data rate of up to 299.6 Mbits/second downlink and up to 75.4 Mbits/second on the uplink when the desirable equipment level and MIMO antenna (multiple input and multiple output antennae and antenna array for example 4X4 or 2X2) setup are used. Much of the LTE standard addresses the upgrading of 3G UMTS to what will eventually be a 4G mobile communications technology. To achieve the desired goal, both simplification and improvements were made on the architecture and air interface of the system. The new air interface becomes E-UTRA or Evolved UMTS Terrestrial Radio Access and the new core network becomes EPC or Evolved Packet Core. Supporting the new E-UTRA air interface is the E-UTRAN or evolved UMTS Terrestrial Radio Access Network where are the base stations, called NodeB in UMTS is now called eNB or Evolved Node B.

Since the first commercial launch of LTE network in 2008 in Sweden by the Swedish mobile network operator TeliaSonera, it was a true success story as more and more LTE networks are being deployed around the world. It is expected that major network operators will migrate their most densely populated networks initially and remaining networks gradually. Latest report estimates that by 2018, over 4 billion users will have the opportunity to enjoy LTE services everywhere they go. With LTE being a true universally interoperable standard, users, whether are accessing their home network or roaming network, will enjoy the same LTE experience.

1.4.3 WiMAX

Indeed WiMAX or Worldwide Interoperability for Microwave Access got a bit of head start on 4G LTE. Based on the air interface defined in IEEE's 802.16 and initially envisioned as a solution for the last-mile in fixed wireless broadband, WiMAX was gaining traction quickly when a group of vendors, suppliers, and operators got together and formed the WiMAX Forum. WiMAX Forum worked furiously to roll out the WiMAX specifications and operators quickly began to deploy network based on the WiMAX specifications. Though ITU initially viewed WiMAX as a 3G technology when the original data rate was designed to provide between 30 to 40 Mbits/second, but the system was easily upgradeable to up to 1 Gbits/second with improvements made on the air interface based on improvements over IEEE 802.16 air interface. Since the original intention of the WiMAX Forum was to define a specification that can deliver last-mile wireless broadband access, many of the distinct features that were present in a true mobile network were not parts of the original specification, such as mobility support, among other things. When ITU defined and published 4G requirements, WiMAX soon followed, but not before many operators had already deployed various versions of networks based on WiMAX on smaller scales when compared to other 3G and/or 4G network rollouts. The first such deployment was as early as 2006 in South Korea and the deployments worldwide have reached over 700 operators. Because of the lower network deployment cost and its incompatibility with other existing network access technology such as CDMA or GSM, many of the deployments were in less developed countries and by new operators who do not have an existing network. In addition, because of its ability as a last-mile wireless broadband solution, many fixed network operator also chose to deploy WiMAX as alternative for fixed network providing broadband access.

1.5 Femtocells

Femtocells were brewed out of the necessity by operators' desire for increase network coverage and faster deployment of networks. Vendors also wanted to shrink the size of the base stations and reduce the cost associated with infrastructure equipment. The concept of Femtocells works for all access technologies from 1G to 4G. Femtocells take on many shapes and forms and are manufactured by various vendors and marketed by even more operators. Some examples are pictured here:

Figure 1.2. Examples of Femtocells

Essentially, Femtocells are small base stations that are originally designed to be deployed at customer premise. For example, a user that works in the city and lives in the suburbs wishes to maintain coverage even when he is at home. Unfortunately, because of the cost of extending wireless networks to the suburbs, the operators tend to extend only when they feel that the user base is big enough to justify the network equipment investment and this limits the coverage rollout of the newer generation of networks in both 3G and 4G. Simply said, an operator will not deploy or extend its network for a few users or will take a long time when the user base is saturated. But with Femtocells, the operator is able to extend its network with a fraction of the cost of deploying traditional networks. Though lacking cellular coverage in many rural areas, but the broadband coverage for Internet access is abundant, even in the suburbs where user may have multitude of options for broadband, such as satellite, cable services, DSL services, or even fiber Internet access services. The Femtocell is a base station, but instead of dedicated backhaul to connect to the operator network, it uses the broadband as the backhaul to route the signals back to the operator's network. The traditional base stations, called macro base stations, cost tens and thousands of dollars where the Femto cells or home base stations only cost a fraction, in the order of one hundred dollar or less. There was even anticipation that as the number of Femtocells grow, the cost of each may come down to as little as $50 due to the economy of scale.

As the Femtocells mature and becoming increasing used by operators to extend their networks, there have been many iterations of the Femtocell, built for accommodating different access technologies, such as the CDMA2000 version, WiMAX version and 3GPP versions. For example, CDMA2000 version of the Femtocells is designed for networks that use the CDMA2000 access technology, such as Sprint and Verizon in the United States or China Telecom in China. 3GPP networks, including UMTS and LTE networks, have their own version of the Femtocells called the 3G or LTE Femto.

This book is organized such that the initial focus is on the Femtocells that are based on the 3GPP technology (both UMTS and LTE), in particular, on the security aspects of Femtocells in communications and in networking. In later chapters, Femtocells from other access technology will be looked in more details. Since much have been written on the various topics of 3G or 4G technologies (UMTS, CDMA2000, LTE, and WiMAX technologies) covering architecture, operations, and security etc., therefore many details about their architectures and operations are left out of this book and are only included (key elements, features and network elements) to aid the understanding of the Femtocells and their operations. Security overview provided in later chapters on UMTS and on LTE would augment reader's knowledge and understanding of the concepts and server as a tool to further grasp the security of the Femtocells.

2

3GPP Architecture and the Femtocell

Before getting into the details of the Femtocell Architecture, it pays dividend to look at the 3GPP network architecture (both UMTS and LTE) briefly and understand how and where the Femtocell fit into the 3GPP network architecture since the Femtocells in 3GPP were primarily designed for UMTS and LTE networks. The 3GPP networks had evolved from the second generation GSM network. First, let us start with the UMTS architecture.

2.1 UMTS Architecture

A UMTS network consists of three interacting domains; Core Network (CN), UMTS Terrestrial Radio Access Network (UTRAN) and User Equipment.

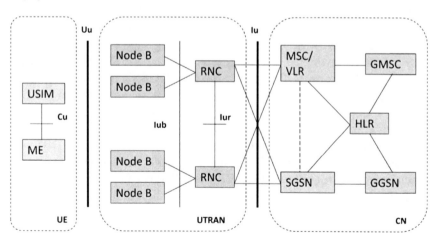

Figure 2.1 UMTS Network Architecture and Network Interfaces

The main function of the core network is to provide switching, routing and transit for user traffic. Core network also contains the databases and network management functions.

The basic Core Network architecture for UMTS is based on GSM network as well as GPRS network that evolved from GSM. Since GSM and UMTS had different air interfaces, network interface and network elements, major equipment modification and enhancements were needed for UMTS operation and services.

The UTRAN provides the air interface access method for User Equipment, which still uses a SIM-card based removable user identity module now called USIM. Base station is referred to as Node-B and the Radio Network Controller (RNC) houses many NodeB functions and interfaces to the core network. The architecture of the network was becoming flatter.

UMTS still needed and supported circuit-switched (CS) network to carry the voice traffic in addition to the packet-switched (PS) network for the non-voice data traffic and are separated by CS domain and PS domain respectively. Some of the circuit switched elements are Mobile services Switching Center (MSC), Visitor Location Register (VLR) and Gateway MSC. Packet switched elements are Serving GPRS Support Node (SGSN) and Gateway GPRS Support Node (GGSN). In addition to these dedicated elements, EIR, HLR, HLR, AuC are shared between the two different domains. To support additional services and features, other network elements (not shown in the Figure 2.1) can be introduced into the architecture, such as Number Portability Database (NPDB) and Gateway Location Register (GLR for handling of subscribers at network boundaries). Furthermore, MSC, VLR and SGSN can merge to become a UMTS MSC for simplicity in terms of operation and deployment calls for both a new UMTS network and a GSM overlay network at the same time.

The air interface of UMTS is essentially wide band CDMA technology that is a direct sequence CDMA system where user data is multiplied with quasi-random bits derived from the wide band CDMA spread codes. Both frequency division duplex (FDD) and time division duplex (TDD) are used.

The main functions of a NodeB are:

- Air interface Transmission / Reception
- Modulation / Demodulation
- CDMA Physical Channel coding
- Micro Diversity
- Error Handing
- Closed loop power control

The main functions of the RNC are:

- Radio Resource Control
- Admission Control
- Channel Allocation
- Power Control Settings
- Handover Control
- Macro Diversity
- Ciphering
- Segmentation / Reassembly
- Broadcast Signaling
- Open Loop Power Control

UE consists of USIM and ME and has an air interface that interacts with the air interface of the NodeB. Many of the UMTS identities are carried over from GSM days, include:

- International Mobile Subscriber Identity (IMSI)
- Temporary Mobile Subscriber Identity (TMSI)
- Packet Temporary Mobile Subscriber Identity (P-TMSI)
- Temporary Logical Link Identity (TLLI)
- Mobile station ISDN (MSISDN)
- International Mobile Station Equipment Identity (IMEI)
- International Mobile Station Equipment Identity and Software Number (IMEISV)

The UMTS SIM card was improved over the GSM SIM card as well and has several distinct functions that were not part of the GSM SIM card:

- Support of one or more User Service Identity Module (USIM) application(s)
- Support of one or more user profile on the USIM
- Support the ability to update USIM specific information over the air right there on the card
- Security functions such as key generation
- User authentication (i.e. Authentication and Key Agreement functions)
- Optional inclusion of payment methods
- Optional secure downloading of new applications
- Support of extended phonebook

2.2 LTE

Similar to UMTS architecture, LTE architecture also consists three domains: Evolved Packet Core (EPC), Evolved UMTS Terrestrial Radio Access Network (E-UTRAN) and User Equipment.

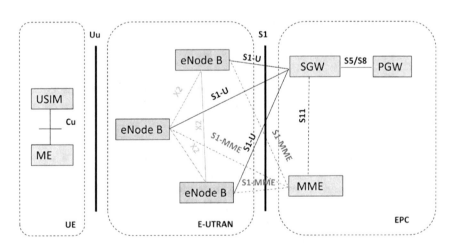

Figure 2.2. LTE Architecture and Interface

LTE did away with circuit-switch domain altogether and relies on packet-switch for everything, including the all-IP packet core called the Evolved Packet Core or EPC. The EPC communicates with packet data networks, the Internet and IP multimedia subsystems. Not shown in the EPC are HSS (Home Subscriber Server), PCRF (Policy Control and

Charging Rules Function), databases, network management functions and few other functional elements that are essentially carried over from UMTS and/or GSM. The main function of the EPC is to provide switching, routing, and transit for user traffic and security for signaling traffic. Below is an oversimplified description of the major components in the EPC:

- Serving GW functions

- PDN GW functions include:
    - Policy Enforcement (QoS, charging, mobility)
    - Per-user based packet filtering
    - Mobility anchoring for intra- and inter-3GPP mobility (requires GTP and MIP HA)
    - Charging Support
    - Lawful Interception

- The Home Subscriber Server (HSS) component has been carried forward from UMTS and GSM and is a central database that contains information about all the network operator's subscribers.

- The Packet Data Network (PDN) Gateway (P-GW) communicates with the outside world (i.e. packet data networks PDN), using SGi interface. Each packet data network is identified by an access point name (APN). The PDN gateway has the same role as the GPRS support node (GGSN) and the serving GPRS support node (SGSN) with UMTS and GSM. It has policy enforcement function (QoS, charging, mobility) and acts as mobility anchoring for intra- and inter-3GPP mobility events.

- The serving gateway (S-GW) acts as a router, and forwards data between the base station and the PDN gateway. It is also the Local Mobility Anchor point for inter-eNodeB handover (i.e. GTP termination), provides PMIP or GTP support towards PDN Gateway. Lawful Interception and traffic accounting are also supported

- The mobility management entity (MME) controls the high-level operation of the mobile by means of signaling messages and Home Subscriber Server (HSS).

- The Policy Control and Charging Rules Function (PCRF) is a component which is not shown in the above diagram but it is responsible for policy control decision-making, as well as for controlling the flow-based charging functionalities in the Policy Control Enforcement Function (PCEF), which resides in the P-GW.

The E-UTRAN handles the radio communications between the mobile station and the evolved packet core and has just one component, the evolved base stations. The evolved base station is referred to as Evolved Node-B, simply as the eNodeB or eNB for short. An eNB controls the mobile stations in one or more cells as well as interfaces with the EPC. The serving base station that is directly communicating with a mobile station is known as its serving eNB. Each LTE UE communicates with just one eNB and one cell at a time (except in later version of LTE called LTE-Advanced which may support multiple air interfaces in carrier aggregation mode to support higher data rate) and the main functions supported by eNB are as follows:

- Sending and receiving radio transmissions to all the mobiles using the analogue and digital signal processing functions of the LTE air interface.

- Implementing control functions for lower-layer operations for all of mobiles supported

- Terminating S1-U interface from the S-GW and S1-MME interface from the MME

- Performing security functions at the PDCP layer (e.g. ciphering and deciphering of user plane traffic)

- Performing intercell radio resource management functions

- Responsible for eNB measurement, configuration, and provisioning functions

- Responsible for other radio related functions such as bearer control, admission, resource allocation and scheduling

Each eNB connects with the EPC by means of the S1 interface (S1-U for user plane traffic and S1-MME for control plane traffic) and it can also be connected to nearby base stations by the X2 interface, which is mainly used for signaling and packet forwarding during handover. If X2 interface is not available between two adjacent eNBs, handover of mobile stations that move from one eNB toward the adjacent eNB would have to traverse all the way via the MME as a way to carry the control signaling for the handover procedure.

The internal architecture of the user equipment for LTE is identical to the ones used in UMTS and GSM which is actually a Mobile Equipment (ME). Using the same architecture greatly simplifies mobile equipment design and improves interoperability as well as user experience. The mobile equipment comprised of the following important modules:

- Mobile Termination (MT): This handles all the communication functions.
- Terminal Equipment (TE): This terminates the data streams.
- Universal Integrated Circuit Card (UICC): This is also known as the SIM card for LTE equipments. It runs an application known as the Universal Subscriber Identity Module (USIM).
  A USIM stores user-specific data very similar to 3G SIM card. This keeps information about the user's phone number, home network identity and security keys etc.

Most of the identities used in LTE are also carried forth from UMTS and GSM (see above on UMTS identities) but so included a few others, for example when Closed Subscriber Group and Home NodeB or Home eNodeB are supported for Femtocells.

Though the UMTS compatible SIM cards are being used for LTE, provision was made to both the authentication and key agreement protocol to support potential larger security key sizes as well as providing cryptographic separation between how AKA keys are used in UMTS and in LTE.

2.3 Femtocells and Femtocell Architecture

The concepts are still the same: the user's mobile station (MS in 3GPP2 and WiMAX context and UE in 3GPP context) still connects to the operator network over the air interface via the base station. Due to the flattening of the network architecture in both UMTS and LTE, the base station have been renamed NodeB and eNodeB respectively to reflect the fact that more of the network functions are moving closer to the network as a result. From the network architecture's perspective, there is much overlap Femto architecture for both UMTS and LTE. Both architectures will be discussed together with additional details and emphasis on the differences. The Femtocells in 3GPP are called Home NodeB and Home eNodeB respectively, or HNB and HeNB. When there is no functional difference to be clarified as needed, H(e)NB is also used to denote that either a HNB or a HeNB is being referred to in the context of the architecture and operations.

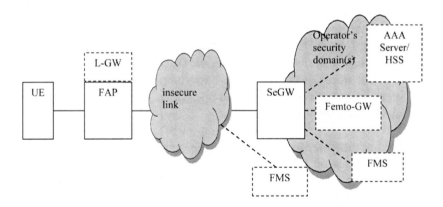

Figure 2.3. 3GPP Femtocell Architecture

In order to use the same terminology throughout the book, it is necessary to make the following clarification:

- the term "FAP" is used to denote any Femtocell or Femto Access Point regardless of the underlying access technology used to implement the Femtocell
- the term "FMS" is used to denote any Femto Management Server, including the Home NodeB Management Server or the Home eNodeB Management Server used in 3GPP exclusively

- the term "Femto Gateway" is used to denote any Femto Gateway server, including the HNB-GW or the HeNB-GW used in 3GPP

From both architectural and security perspective, the Femtocell network is composed of the following network or user equipment and/or functionalities, reference points:

- UE (e.g. MS)
- L-GW (optional)
- FAP
- Internet connection
- Security Gateway
- FMS
- Femto-GW(optional)
- AAA/HLR(optional)
- Iu and S1 interfaces

Description of the system architecture:

- Air interface between UE and FAP in this architecture are compatible to the air interface in UMTS network or LTE network, meaning that a user does not need any specialized equipment other than a mobile telephone that is compatible to GSM, UMTS, or LTE to make and receive calls through the FAP.

- Security Gateway (SeGW) is the door into the operator core network because the signalling required to connect to the operator network are expected to be carried on the public Internet network. There may be some instances where the "insecure link" belongs to the operator in which case the operator really does not consider that as being "insecure", but for the most part, it is just what is – insecure because it traverses through the public Internet.

- An optional AAA server serves as the authentication function for earlier FAPs that uses a separate UICC-based authentication (also called hosting party authentication in the 3GPP Femtocell architecture) to authenticate the FAP to the operator network.

- Once the FAP is authenticated mutually authenticated with the Security Gateway, a secure tunnel is established between the FAP and Security Gateway to protect traffic exchanges, such as control plane traffic, management plane traffic, or user plane traffic. Depending on the content of the traffic, the termination point may be MME for control plane traffic, FMS for management plane traffic, and SGW/PWG for user plane traffic respectively.

- Depending on the deployment scenario, a Femto-GW may be deployed separately or may be integrated with other network entities, such as the Security Gateway. One of its purposes is to perform UE access control. Access control is used as a mechanism to allow or disallow UEs to connect to a particular FAP, since the FAP is being deployed in a customer's premise and there is associated cost involved from the customer's perspective in terms of operating the FAP. The customer may choose not to allow users that the customer does not know or does not want to connect to the FAP located at the customer's premise by forming a Closed Subscriber Group (CSG) white list that contains only the UEs that are allowed to access the FAP (e.g. receive and making phone calls). This CSG capability may be implemented in the network, the UE, or none at all for the UMTS version of the Femtocell. Therefore, UE access control may be performed either at the Femto-GW or at the FAP, depending on whether the UE or FAP supports this CSG capability. At either case, the interface between the SeGW and the Femto-GW and between the Femto-GW and MSC/SGSN are protected using 3GPP defined network domain security mechanisms (i.e. IPsec-based).

- UE access control is needed where the FAP performs optional UE access control in the case of non-CSG capable UEs or non-CSG capable FAPs. SeGW and Femto-GW are logically separate entities within operator's network, with the SeGW located in front of the Femto-GW.

- In the LTE case, Femto-GW is optional to deploy and may be integrated with SeGW also, much like the Femto-GW for UMTS. Even with the SeGW sitting at the edge of the operator network that acts as a single point of entry into operator network, the links behind the SeGW (e.g. interface between SeGW and MME/S-GW, the interface between SeGW and Femto-GW, interface between Femto-GW and MME/S-GW) may require additional protection using IPsec-based protection mechanism as defined in 3GPP's network domain security specifications. Any use of such additional protection is strictly based on operator policies.

- FMS performs location verification of the FAP. Location verification is an important aspect of operating Femtocells that can relieve the operator legal liabilities by enforcing the FAPs are only allowed to operate (i.e. transmit and receive via air interface) where the operator holds legitimate license. Another important aspect of location verification is to prevent unscrupulous users from accessing a particular operator's network without incurring legitimate charges or be charged with a higher tariff (e.g. roaming charges). For example, if a user takes a Femto cell from an operator in country A (e.g. France) to country B (e.g. China) and calls home from country B through the Femto, the operator not only violates the spectrum licensing agreement because it may or may not have the license to operate in country B, but it also suffers revenue loss because from the operator's perspective, the call made or received by the user would have been considered as originated or terminated in country A. Another point about location verification is for the benefit of the user in case of emergency access where the operator can correctly route the call to the emergency handling center if the operator knows where the call originated so that proper emergency assistance can be dispatched without delay, potentially saving valuable time especially in life-threatening situations. In the UMTS network, location may also be performed by FAP.

A local gateway (L-GW) is optional to deploy for networks that support local breakout capabilities. Local breakout is a function that allows the user's Internet traffic to go directly between the UE and the Internet without having to go through the operator's core network node (e.g. Serving Gateway) and greatly improving the operator's network's throughput. This is analogous to a user connecting his mobile phone to a Wi-Fi network and uses the Internet or accesses other network nodes that are connected locally, except in this case the user connects to the FAP and access the Internet or other network nodes locally through a local breakout function supported by the L-GW. In case the L-GW is deployed, the signaling still needs to go to the operator's core network via secure connection to the Security Gateway even though the L-GW's Internet traffic goes through to the Internet via a separate Internet connection. When the L-GW is deployed independent of the FAP, the interface between the FAP and the L-GW needs to be secured.

Part II

Securing the Femtocells

3

Security of Femtocells

Through the Femtocell, the operator is able to provide new services with higher data rate at relatively lower cost compared to traditional cellular systems allowing the operator to rapidly provide and expand coverage. However, because Femto Access Point is located in the customer premise and will access the operator's core network via the insecure IP link that the operator generally does not trust, it potentially exposes the operator's core network to threats that the operator had never had to consider. The operator's core network, which is considered the bread and butter of the operator's cellular network, is the key to ensure that not only the smooth operation, but also secure operation of the operator's networks. With the introduction of Femtocells into the operator's network, the secure operator's core network somehow finds itself being accessible via IP-based connection from the wide open Internet, through the interfaces of the Femto access point. The core network can no longer be considered secure, especially with a plethora of potential attacks possible that originated from every corner and crevice of the Internet. An attacker can remotely launch attacks to the operator's core network from anywhere there is an Internet connection (e.g. from the comfort of the attacker's home). The risk is so great that the wireless community spent a considerable amount of time analyzing the security threats, looking at every scenario, and coming up with solutions to prevent any potential threats that the operator's core network could face. Because Femtocells are so different in terms of the threat and security model compared to traditional base stations, the steps that go into securing not only the Femtocell but also the operator's core network go way beyond that of the either UMTS or LTE alone.

Before dwelling deeper in the realm of Femto security, it may be beneficial to look at the UMTS and LTE security models briefly for better understanding.

3.1 UMTS and LTE Security

3GPP has taken evolutionary steps in terms of security improvements over GSM in UMTS and over UMTS in LTE to address the shortcomings from the previous versions. The security improvements are also influenced by advances in cryptographic algorithm design, security protocol design, and equipment capability.

Years of GSM operations exposed some of the critical security shortcomings of the GSM technology and the network. GSM was designed to provide only access security over the air interface while the backhaul was not protected. It assumed that the backhaul was sufficiently secured by physical means and that the backhaul network is a relatively closed network inaccessible by attackers. GSM did not address active attacks in case of base station impersonation scenarios. Due to the limitation of computing capability on the mobile station and on the SIM cards, only 64-bit keys were used for authentication and ciphering. These critical short comings were sufficiently addressed in UMTS security architecture with the following improvements:

- Mutual authentication between UE and base station by way of a new authentication and key agreement protocol (AKA) that ties the generation of the session keys for integrity and ciphering as a result of a successful authentication
- Message integrity protection for the UE to verify the origin and authenticity of the signaling message from the base station to protect against false base station attacks (e.g. base station impersonation)
- Backhaul security with the addition of MAPSEC and network domain security (using IPsec)
- Wider Security Scope by implementing many base station security functions in the RNC
- Ability to extend security features as required by new threats and services
- Enhanced SIM card (USIM on a UICC) that supported larger root key with a key size of 128 bits
- Newer and stronger cryptographic algorithms for integrity protection and ciphering, including Snow 3G, and AES
- Support of variety of integrity protection and ciphering options based on operator policy

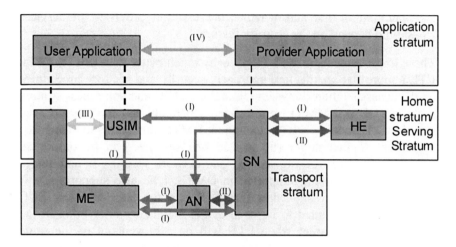

Figure 3.1. LTE Security Architecture

When LTE was in its infancy stages, designers took yet another approach to further improve security that was already in UMTS. The new LTE security architecture defined five separate security feature groups or domains (see above figure), namely:

- Network access security (I): the set of security features that provide users with secure access to services, and which in particular protect against attacks on the (radio) access link.

- Network domain security (II): the set of security features that enable nodes to securely exchange signalling data, user data (between Access Network and the service network and within the access network), and protect against attacks on the wire line or backhaul network.

- User domain security (III): the set of security features that secure access to mobile stations.

- Application domain security (IV): the set of security features that enable applications in the user and in the provider domain to securely exchange messages.

- Visibility and configurability of security (V): the set of features that enables the user to inform himself whether a security feature is in operation or not and whether the use and provision of services should depend on the security feature.

These improvements made LTE security much better than UMTS. With LTE's more flat architecture approach, security and security procedures on the access stratum (between UE and the eNB) and non access stratum (between UE and MME) were defined separately that further improved security. Essentially, LTE extended UMTS authentication and key agreement so that larger (up to 256 bit) key size can be supported, offered better interworking security and network security. In addition to the security algorithms supported in UMTS, LTE also supports a new regional encryption algorithm (i.e. ZUC) in the later releases. ZUC algorithm was designed to be used as the mandatory encryption in the domestic China market when LTE will be deployed out of the same concern that China was too depending on technology developed in other parts of the world.

3.2 Femtocell Security

3.2.1 Femtocell Threats and Attack Model

Designing of the Femtocell security took the same approach as in LTE and UMTS in 3GPP. By looking at the threats and attack models, it would become clear what the risks are and what assets need to be protected. The security architecture of the (3GPP) Femtocell looks strikingly similar to that of the (3GPP) LTE for good reasons. The remaining sections within this chapter will be devoted to the Femtocell based on the 3GPP UMTS and LTE. As mentioned previously, FAP and the 3GPP equivalent HNB, HeNB or H(e)NB are used interchangeably throughout this chapter and later chapters to denote a Femto Access Point without distinction, especially in the 3GPP context of things.

In the early stages of Femtocell security design several years ago, the security group within the 3GPP's standards producing organization performed a comprehensive feasibility study on the security aspect of Femtocell from various aspect of the system and identified a complete set of almost thirty potential threats and attacks that are can be applied to

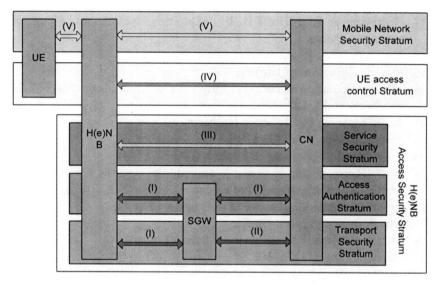

Figure 3.2. Overview of (3GPP) Femtocell Security Architecture

one or more assets, or components, in the Femtocell and these threats were captured in a 3GPP technical report. These threats as listed below without ranking the seriousness of them:

1. Compromise of H(e)NB authentication token by a brute force attack via a weak authentication algorithm.
2. Compromise of H(e)NB authentication token by local physical intrusion.
3. Inserting valid authentication token into a manipulated H(e)NB.
4. User cloning the H(e)NB authentication Token.
5. Man-in-the-middle attacks on H(e)NB first network access.
6. Booting H(e)NB with fraudulent software ("re-flashing").
7. Fraudulent software update / configuration changes.
8. Physical tampering with H(e)NB.
9. Eavesdropping of the other user's UTRAN or E-UTRAN user data.
10. Masquerade as other users.
11. Changing of the H(e)NB location without reporting.
12. Software simulation of H(e)NB.
13. Traffic tunneling between H(e)NBs.
14. Mis-configuration of the firewall in the modem/router.
15. Denial of service attacks against H(e)NB.
16. Denial of service attacks against core network.

17. Compromise of a H(e)NB by exploiting weaknesses of active network services
18. User's network ID revealed to H(e)NB owner
19. Mis-configuration of H(e)NB
20. Mis-configuration of access control list (ACL) or compromise of the access control list
21. Radio resource management tampering
22. Masquerade as a valid H(e)NB
23. Provide radio access service over a CSG
24. H(e)NB announcing incorrect location to the network
25. Manipulation of external time source
26. Environmental/side channel attacks against H(e)NB
27. Attack on OAM and its traffic
28. Threat of H(e)NB connectivity to network access
29. Handover to CSG H(e)NB.
30. H(e)NB presenting different security credentials and identities to H(e)NB and H(e)NB-GW respectively

The serious nature and potential impact of these identified threats varies depending on the particular threat. Not every threat may result into a direct attack by itself while some attacks may results in multiple threats being realized. Threats and attacks, when combined together may present additional challenges for the designers and implementers. Additionally, each threat impacts one or more assets. The assets not only can be in the form of physical asset such as Femtocell, Femtocell Gateway, Security Gateway or other network component but also stakeholders such as the users and the operators as can be seen in Table 3.1.

A threat by itself, if not exploited, remains simply a threat without physical harm to an asset or part of a system. On the other hand, attacks are achieved by means of one or more threat being exploited and realized. In order to understand how the threat is exploited, the potential attack that can be associated with the threat is described in detail, along with how the attack impacts the assets being protected. At the same time, some mitigation techniques are also suggested that can be used to counter the threats and attack identified. Further security mechanisms will be described in later sections. Note that a particular security mechanism may be applied to counter one or more threats as some of the threats cause similar impact to the system.

Table 3.1. Threat-Asset Corresponding Table

Threat/Asset correspondence	H(e)NB	User	Operator
Threat-1	X	X	X
Threat-2	X	X	X
Threat-3	X	X	X
Threat-4	–	X	X
Threat-5	–	X	X
Threat-6	X	X	X
Threat-7	X	X	X
Threat-8	X	X	X
Threat-9	X	X	–
Threat-10	X	X	–
Threat-11	X	X	X
Threat-12	X	X	X
Threat-13	–	X	X
Threat-14	–	X	–
Threat-15	–	X	–
Threat-16	–	X	X
Threat-17	X	X	X
Threat-18	–	X	–
Threat-19	X	X	X
Threat-20	X	X	X
Threat-21	X	X	X
Threat-22	–	X	X
Threat-23	–	X	X
Threat-24	X	X	X
Threat-25	X	X	X
Threat-26	X	X	X
Threat-27	X	X	X
Threat-28	–	X	X
Threat-29	–	X	X
Threat-30	X	X	X

An attacker may launch an attack from a spoofed network access concentrator on the internet. When a weak or broken authentication or encryption algorithm is used or even when a weak key is used, it makes the attacker's job so much easier. Encryption algorithms are not designed with weak keys in mind, but many algorithms have been found with weak keys. Even algorithms with weak keys may still be effective as long as the keys used in the algorithms are checked with a known list of weak keys before applying the session keys. Some other algorithms are completely broken. Example of such an algorithm is the GSM A5/2 algorithm or COMP128-1, which was found to have weakness and subsequently found to be broken, has been removed from the mobile systems after a long debate in the 3GPP community. The removal of such an algorithm was both painful and lengthy as it involved the collaboration among operators, mobile equipment manufacturers and handset vendors. When the algorithm is symmetric using pre-shared key, If a weak key is used, traditional brute force attack is possible and such attack is made even easier as the computational power of the current generation of computers and/or distributed nature of networked attack methods are becoming more prevalent. An example of a weak key is one that when used with a specific ciphering algorithm makes the cipher behave in some undesirable way, leading to the easier recovering of plaintext encrypted with the cipher and/or even the key that is used to encrypt the plaintext. This attack and other attacks may lead to an attacker gaining unauthorized access to H(e)NB, using the H(e)NB as an attacking tool to launch further attacker to other H(e)NB, operator network or other users.

Mitigation: Any authentication token with a weak algorithm like GSM SIM with COMP128-1 should not be used for H(e)NB authentication. Backhaul link protection mechanism should be strong enough.

2) Compromise of H(e)NB authentication token by local physical intrusion

Since the H(e)NB is designed to be located within the customer's premise, the attacker has the full H(e)NB at his disposal and at his convenience. The attacker may physically open the H(e)NB or tap into the wiring of the H(e)NB in an attempt to recover the stored authentication token. Once the authentication token is recovered, the attacker then can take a copy of that and install it in another H(e)NB in an impersonation attack. This type of threats is highly likely if the H(e)NB is implemented in general computing platform without any hardware-based security, such as Trusted Platform Module that is commonly found in many portable personal computers today or if the authentication token is not stored in a UICC, a device that is universally used in all 3GPP mobiles as a secure device for authentication, key storage, etc.

Mitigation: Authentication credentials of the H(e)NB shall be stored inside a secure domain i.e. from which outsider cannot retrieve the credentials.

3) Inserting a valid authentication token into a manipulated H(e)NB.

GSM, UMTS and LTE using security tokens or credentials stored on the SIM cards (e.g. UICC) that has made provisioning of shared secret between user equipment and the network simple and transparent where a user would simply plug the SIM card in any mobile phone and can start accessing the network. When the SIM cards are sent to the operator for distributing to new users (or users whose SIM is damaged or otherwise needing replacement), the security tokens are also stored in the operator's database, ready to be used. The advantage of using UICC for service provisioning of user mobile equipment and the ability to allow users to choose any user mobile equipment they wish is a disadvantage in the network equipment model, especially when the network equipment is easily accessible to the attacker (i.e. installed at customer premise). If a removable token is used, the attacker may remove it and use it in another unauthorized H(e)NB to access the network. Example of unauthorized H(e)NBs are H(e)NBs that are modified to inflict damage to the network or even as simple as a H(e)NB that is designed for Operator A but put to use in Operator B's network. Consequences of such an attack may introduce malicious configuration changes to the H(e)NB and resulting in compromise of user signaling and data, H(e)NB transmitting in unauthorized spectrum or at an unauthorized or unacceptable power level causing interference of other H(e)NBs, or even H(e)NB impersonating another legitimate H(e)NB.

Mitigation: Avoid using removable token as a way to store security credentials or require a cryptographic binding between the removable token and the H(e)NB.

A non-removable authentication token is helpful to mitigate the risk. Also new users could be required to explicitly confirm their acceptance before being joined to a H(e)NB. This way a H(e)NB owner could only perform eavesdropping/masquerade attacks against those who join the H(e)NB. This approach relies on additional access control being enforced in core network, not just only at the H(e)NB.

It is possible that introducing device authentication or binding removable token to certain H(e)NB can also mitigate the risk, which may need a combination of a removable token and an onboard token.

4) User cloning the H(e)NB authentication Token.

This attack requires that the token used to authenticate H(e)NB can be (easily) cloned and the cloned token is then inserted in a genuine H(e)NB.

As discussed above, one potential threat of compromised security token is that the attacker can take the exposed security token and clone it to be used in other H(e)NBs. Cloning SIM cards are especially easy with various SIM cloning tools readily available in the market place, especially for the older SIM cards used in GSM mobile stations. However, the newer generation of SIM cards (e.g. 3G UICC) are more difficult to clone, but still could not prevent the determined attacker with the necessary know-how and tools to do it. Once the security token is cloned, it is used by the attacker can use it in a legitimate H(e)NB and launch other potential attacks. In this case, there is no direct threat to the H(e)NB itself, but the threats to users and operators are immensely grave. Since security tokens in 3GPP networks are used to further derive ciphering keys, the attacker gains the ability to eavesdrop all communication going through the H(e)NB. Depending on the mode of operation of the H(e)NB, whether it is open which means all UEs can connect without having to go through CSG verification or whether it is closed which means only UEs in the CSG can connect, the damages to user's privacy and operator's reputation could be devastating. Additionally, the operator may not be able to generate billing records correctly for UEs that are connected through the H(e)NB, results in further revenue loss.

Mitigation: the authentication credentials of the H(e)NB should be difficult to clone. Also new users could be required to explicitly confirm their acceptance before being joined to an\ H(e)NB. This way a H(e)NB owner can only perform eavesdropping/masquerade attacks against those who join the H(e)NB. This approach relies on additional access control being enforced in core network, not just at the H(e)NB. Multiple instances of the same H(e)NB should not be allowed simultaneous access to the core network. Some forms of location locking (e.g. to DSL line) may also help to mitigate this threat.

5) Man-in-the-middle attacks on H(e)NB first network access

This threat is possible when the H(e)NB does not have any unique authentication credentials pre-installed at the factory or inserted into the H(e)NB, similar to some user equipments used in IS-95 CDMA networks that rely on ephemeral Diffie-Hellman key exchange protocols to establish pre-shared secret over the air as part of over-the-air service provisioning procedures. It was shown that even with a weak password the password authenticated Diffie-Hellman key exchange protocol can still provide good protection against man-in-the-middle attacks. In case of H(e)NB, the Diffie-Hellman key exchange does not necessarily have to take place over the air, it can also be conducted over wired connection to make it more difficult for the man-in-the-middle to initiate an attack.. However, without password authentication, a man-in-the-middle attacker can still jump in the instant that the H(e)NB initiates an ephemeral key exchange with the network by establishing a key exchange session with the H(e)NB and another key exchange with the operator's network in both wired and wireless scenarios. This attack is possible because the operator's network cannot reliably identify the H(e)NB that initiated the key exchange. As a result, a man-in-the-middle attacker may intercept all traffic originated and terminated on the H(e)NB, compromise users' privacy, and impersonate the H(e)NB.

Mitigation: H(e)NB shall have authentication credentials already during the very first contact with the network. These credentials shall be recognized at the operator's side. Un-authenticated traffic should not be accepted even at the "first-contact" phase. Either a USIM on a UICC, or a vendor certificates can be used for this purpose. But the logistical consequences could be different. UICC could be inserted in the H(e)NB at the point of sales or customer. Vendor certificate has to be inserted in the H(e)NB at stage of manufacturing. If key exchange over wireless or wired connection is used, authenticated key exchange similar to the optional password authenticated Diffie-Hellman key exchange protocol defined in CDMA2000's C.S0016 or in the IETF RFC 5683 is more secure. Depends on the type of credential and where the credential is stored, the end point where authentication is performed need to be clearly identified. For example, if device certificate is used, mutual authentication is performed between the nodes where the certificates are provisioned (i.e. Security GW and H(e)NB). If UICC is used in storing the security credentials, mutual authentication is to be performed between HSS and UICC.

6) Booting H(e)NB with fraudulent software ("re-flashing")

This threat is possible given the easy access of the H(e)NB installed in the customer premise and the amount of information available on the Internet. An attacker with physical access of the H(e)NB and the technical know-how may try to reverse-engineer or modify the software in the H(e)NB and to gain certain advantage of the H(e)NB, similar to re-flashing software or firmware in certain routers or even certain mobile phones to avoid restrictions or to unlock more advanced features. The threats are to both H(e)NB and the operator's core network are of great concern with potentially devastating attacks ranging from eavesdropping on user communications, impersonation towards the network, disruption of operation of the operator's network, and even denial of service.

Mitigation: Booting process shall be secured by the cryptographic means, for example using a TPM module. Additional security measures may be needed in case of USIM-based H(e)NB authentication towards the network.

7) Fraudulent software update / configuration changes

Fraudulent software update and configuration changes would also cause similar effects as the user running or booting from modified software with similar results.

Mitigation: All software updates and configuration changes shall be cryptographically signed, and H(e)NB shall have means to verify the signature.

8) Physical tampering with H(e)NB

When the H(e)NB is within physical reach of potential attackers, the attacker may try to modify or enhance the functionality of the H(e)NB. An example of such would be to replace the H(e)NB's antenna with one that is of higher gain, effectively extending the range of the signals. Since the original design of any Radio Frequency (RF) component of the

H(e)NB has gone through extensive testing (e.g. conformance testing, interference testing, environmental impact, etc.) to comply with various regulations (e.g. FCC regulations) and conformance standards, any modification could easily change and characteristics of the RF components resulting in nullifying the benefits of the extensive resting.

Mitigation: H(e)NB shall be physically secured to a moderate extent to prevent easy replacement of components. Trusted computing techniques could be used to detect when critical components are modified or replaced.

9) Eavesdropping of the other user's UTRAN or E-UTRAN user data

This requires that some parts of the user plane ciphering to be terminated inside the H(e)NB and as a result the H(e)NB leaves user traffic unprotected in some part of the H(e)NB.

If the H(e)NB is allowed to be configured in open access mode by the user, an attacker (i.e. the user) may, either through running modified software or using probe to connect to the internals of the H(e)NB, try to access the unsuspecting user's data. If the user data is either available unprotected on air-interface or with IP-based interface security not implemented, the attacker may succeed in gaining access to user's data, since in the current 3GPP architecture (both UTRAN and E-TRAN), the user plane data protection terminates at the (e)NB and in this case the H(e)NB. Though the tampering and physical tapping of the physical bus and wires of a device is still very difficult, but then again, the attacks have become more sophisticated recently.

Mitigation: Unprotected user data should never leave a secure domain inside the H(e)NB. The user could be notified when the UE camps on a closed or open type H(e)NB. User could be notified (or give his/her explicit acceptance) when he/she is added to the access list of a closed access mode H(e)NB.

10) Masquerade as other users

Similarly to the above threat, once an attacker gains access to the internals of the H(e)NB, the attacker is not only able to listen to the conversation of an user, but he may potentially be able to pretend to be the user, for example, by maintaining the session of the user once the user terminates his session. The session may appear to be terminated from the user's perspective, but the attacker controlled H(e)NB may keep the user session alive without the user's knowledge and authorization and launch other attacks. When the H(e)NB is configured in open access mode, the eavesdropping can happen to any UE that happens to connects to the H(e)NB.

Mitigation: Unprotected user data should never leave a secure domain inside H(e)NB. The user could be notified when the UE camps on a closed or open type H(e)NB. User could be notified (or give his/her explicit acceptance) when he/she is added to the access list of a closed mode H(e)NB.

11) Changing of the H(e)NB location without reporting

One of the operator's responsibilities in operating the spectrum license is to operator it where the spectrum license is authorized to operate. The spectrum is carefully allocated to provide coverage and to avoid interferences among wireless devices. With the traditional wireless network, every piece of network equipment that is RF-capable (e.g. base stations) has been carefully placed into service to maximize its benefit (i.e. providing the best coverage with the least amount of interference). Since every piece of network equipment is well within the control of the operator, the equipment placement is also well with the control of the operator. However, with the H(e)NB to be placed in customer premise needing only an Internet connection and a power receptacle to operate, the user may take the H(e)NB to another location, a location where the operator does not have license to operate its wireless network and provide wireless service, for example, a vacation house or a hotel in another region or country. If the H(e)NB is allowed to operate, the operator will not be able to determine the location of the H(e)NB and will potentially be held responsible for violating the legal requirement of spectrum licensing agreement where the operator obtained the spectrum. In this case, though there is not threats to H(e)NB or user, but the threats to the operator are enormous, ranging from loss of revenue, unable to provide emergency service, unable to provide legal intercept capability if required by regulation, or even fine and other legal penalty from the spectrum licensing agency.

Mitigation: Appropriate location verification mechanism shall be designed and implemented. If a removable token-based approach is used for authenticating the H(e)NB (case 3 or 4), it may be easier for an attacker to benefit from a weak or non-existent location locking mechanism.

12) Software simulation of H(e)NB

Since many of the today's network equipments can be built using available general purpose hardware with, it is also foreseeable that an attacker with the know-how and the necessary skills and equipment may create a simulated version of the H(e)NB that runs on general purpose computing platform. If the simulated software version of the H(e)NB is able to connect to the home network with or without the user's consent, the attacker can cause some damage to either the operator's network, provided that this threats requires very sophisticated skills and deep understanding of operation of the H(e)NB. Unfortunately, most of the literature is available on the Internet for a determined attacker. The side effect of this threat to the H(e)NB is that the operator may not be able to identify whether the H(e)NB in question is a legitimate H(e)NB or one that is simulated resulting in revenue losses. The operator also may not be able to fulfill any legal requirements such as lawful intercept, resulting in further legal ramifications. Furthermore, denial of service attacks could also be carried out by the attacker.

Mitigation: As software simulation cannot be prevented, is it necessary to enforce strong H(e)NB access authentication and to prevent removal/cloning of the authentication token.

13) Traffic tunneling between H(e)NBs

When a legitimate H(e)NB is used at an operator authorized location (i.e. the operator's home network location) but if it is allowed, for example, to connect to another H(e)NB that is not in the operator's authorized location (i.e. where operator does not have spectrum license) through the H(e)NB's modified software or to a H(e)NB that pretends to be SeGW, the connection may allow the attacker to tunnel UE traffic through the H(e)NB at the unauthorized location. This threats may create an overload condition to the legitimate H(e)NB as it now has to support UEs that are connected to it and also channel traffic through a tunnel to an attacker's H(e)NB supporting that H(e)NB's UE traffic. Furthermore, this threat also allows voice and data traffic to originate from any location, in particular, in unauthorized location where the operator is not allowed to operate, resulting in the operator suffering from revenue loss and potential inability to provide lawful intercept and/or emergency services.

Mitigation: The H(e)NB should be able to detect abnormal traffic that does not originate from locally connected UE. One countermeasure is to enforce that only authenticated UE is allowed to be used with the H(e)NB.

14) Mis-configuration of the firewall in the modem

When H(e)NBs are intentionally integrated with Internet devices such as DSL/cable modem, home router, or Wi-Fi access point that have direct access to the Internet, the access and configuration of the firewall capabilities are usually controlled through web interface. Incorrect configuration of the firewall, for example incorrect setting of TCP or UDP ports to communicate with gateway elements in the operator's core network may prevent the H(e)NB from able to connect to the operator's network, especially when users are not skilled in setting up and/or configuring firewalls and access rules of the modem. The side effect of this threat is that the user is not able to configure the H(e)NB correctly and that the H(e)NB is not able to connect to the operator's network, causing the user not able to enjoy services provided by the H(e)NB. Though this threat is perceived as unintentional and there is no attacker in this case, but it can also results in the operator's loss of revenue and loss of time spent in providing the user with technical support and/or on-site visit to correctly configure the settings for the user.

Mitigation: In case when the modem or router is integrated with the H(e)NB, it shall have pre-defined and not changeable configuration of the H(e)NB access channel. In case when the modem is a separate box, its correct configuration shall be enforced. One possible approach may be using uPnP mechanism. An additional firewall within the H(e)NB would also be useful.

15) Denial of service attacks against H(e)NB

This type of threats is perhaps the easiest to launch and attacks are possible on just about every layer of the protocol stack that ranges since majority of the software and protocol stack is modeled after the seven layers of the Internet protocol. Example of ARP or IP related attacks are possible on layers 1 through 3. TCP, IGMP, and UDP based attacks are possible on layer 4. Attacks based on applications and protocols supported by H(e)NB on higher layers are possible on layers 5 through7. Since H(e)NB is also considered an IP-based device, it is no more susceptible to denial of service attacks than other network gear such as routers and/or internet gateways. The threat to H(e)NB is that it is not able to function properly, denying the user the satisfaction of enjoying H(e)NB service.

Mitigation: H(e)NB is partially relieved from the processing load if a firewall at the modem is present, and configured to pass only IKE negotiations and ESP-encrypted traffic to the H(e)NB. We note that IKEv2 (when used on S1 or X2 interface for user plane traffic) is more robust against DoS attacks than IKEv1. When the IP-level cryptographic protection of the S1/Iu-link is used, DoS traffic (which is assumed to be unauthenticated) is filtered out already at the authentication phase.

16) (Distributed) Denial of service attacks against core network

When multiple H(e)NBs are under the control of attackers, potentially through distributed fashion, using any of the attacks on various layers of the H(e)NBs, the attacker may be able to disrupt the entire operator's network. Again, these types of attack are no more susceptible than other Internet devices, but because of the attractiveness of breaking into operator's core network, which has been traditionally considered closed, it is especially enticing for the attackers to launch denial of service attack, causing disruption of service, suffering from outages on the backhaul link and overload of other network elements for the operator costing time and money to remedy. Because H(e)NB and SeGW runs IKEv2-based authentication and IPsec tunnel establishment protocol, the any one of the attacks against IKEv2 protocol is possible and they include IKE_SA_INIT flood attack, IKE_AUTH attack, Flood of legitimate tunnels attack (exhausting resources on the Security

Gateway), Malformed IKE_SA packets, Malformed authentication credentials, etc.

Mitigation: Secure core network elements that shall be secured include security gateway as first contact point in the core network. Securing security gateway also means that for layer 3-7 volume attacks, the Security Gateway shall be remain available in the event that a high rate of IPsec IKEv2 signaling messages are handled by the Security Gateway. The Security Gateway shall protect the upstream network from overload and overflow conditions.

17) Compromise of a H(e)NB by exploiting weaknesses of active network services.

Various network services are needed for the operations of the H(e)NB, such as protocol handlers or port mappers. Remote-based IP attacks have been well documented that have been used by attackers to on network devices. These attacks may cause protocol handlers to fail, and when launched against H(e)NBs, the entire H(e)NB may be compromised. These attacks are made possible over a widely distributed area affecting multiple H(e)NBs, resulting in loss of privacy, impersonation towards the network, and potential service interruptions and threats against operator's core network and its management components.

Mitigation: Minimized network services (disabled or firewalled), robustness testing for functional protocol handlers, intrusion detection looking for abnormal H(e)NB behavior, regular reset to a securely verified system state.

18) User's network ID revealed to H(e)NB owner

To protect a user's privacy, both UMTS and LTE networks use various temporary and pseudo identities. However, when the H(e)NB's owner is able to perform management of the CSG list, for example by adding or deleting a particular user from the CSG list, there is a potential for that particular user's temporary identities to be exposed and linked to the user's true identity. Though this link or binding exists on the network and is known to the network operator, but when it is known to someone that is not part of the operator's network, it subjects a particular user to privacy leaks.

Mitigation: A link between IMSI and owner given user ID is stored in the network or secure stored in the H(e)NB.

19) Mis-configuration of H(e)NB

The correct operation of the H(e)NB depends very much on the internal configuration parameters. Usually, these configurations are coordinated by the operator's H(e)NB Management System (HMS) over secure connections (e.g. IPsec or TLS). However, if the attacker, having physical access to the H(e)NB, is able to make configuration changes, there is some potential for damage. If the configuration modification nets the attacker's ability to forward traffic, the attacker then could remotely record the traffic and recover the content offline (e.g. privacy leak for the user) or even denial of service attack against the operator network by forcing large amount of unwarranted or useless traffic through the backhaul

Mitigation: Secure access to configuration of H(e)NB is needed.

20) Mis-configuration of access control list (ACL) or compromise of the access control list

If the attacker is able to gain access to the access control list (ACL) through administrator's access through physical access, the attack would be able to change the permission of the access by allowing unauthorized UEs to bar disallowed UEs from accessing the H(e)NB. The threats of this is that users may be denied access (e.g. DoS attack on the users) or the operator may suffer revenue loss, especially if the there is flat billing based on H(e)NB.

Mitigation: Secure means of creation, maintenance and storage of ACL is required.

21) Radio resource management tampering

One of the important aspects of operating the H(e)NB or any eNB in a wireless network is compliant with regulations governing the transmission power of the devices. If, for example, the configuration parameters governing the transmission power are modified, any part of the power control components are tampered with or even range extension/signal booster devices (e.g. external antenna) are installed, the H(e)NB could be transmitting at a higher power level than that is allowed by regulatory requirements and cause interference to neighboring H(e)NBs or macro (e)NBs or even increase in the number of undesired handovers between H(e)NBs and (e)NBs. This would cause degradation of the network performance for the operator as the operator network RF characteristics have been carefully calibrated based a number of factors such as terrain, aerial topology, distances between (e)NBs, signal strength, etc.

Mitigation: There should be no means to control the radio resource related parameters by a user. The configuration interface of the H(e)NB must have adequate security. It will be difficult to provide protection against range extension.

22) Masquerade as a valid H(e)NB

This has been identified as a potential threats to users and the operator network. The attacker, using either a H(e)NB purchased off-the-shelf or one that is simulated by software, is able to configure the H(e)NB to operate in such a way that it allows users of a particular CSG to join or to operate in open access mode. Additionally, the attacker also reconfigured the H(e)NB to disable integrity protection and confidentiality protection. By doing this, the attacker is able to compromise the privacy of users who connect to the H(e)NB and aid certain users to evade charging.

Mitigation: CSG setting and other configuration should be hidden. There should be binding between H(e)NBs and the users it can serve that should also be known by the network. The H(e)NB must be authenticated by the network. The case of key leakage requires that the keys in a H(e)NB is stored in a secure location.

23) Provide radio access service over a CSG

In this threat scenario, the attacker has a legitimate H(e)NB and has valid connectivity to a CSG. Since the attacker is in physical possession of a

H(e)NB, the attacker can use one of several means to gain access to eh CSG, for example, connecting to the H(e)NB using Ethernet cable, connecting to the H(e)NB as a UE or connecting to the H(e)NB through a Wi-Fi access point using a laptop. Through this attack, denial of service to the operator network is possible, user loss of privacy is possible, and operator revenue loss is also possible.

Mitigation: Radio layer forwarding is difficult to mitigate. They might require RF fingerprinting. Network layer forwarding attacks require similar mitigation as the following threat.

24) H(e)NB announcing incorrect location to the network

By reporting incorrect location to the network, a stolen H(e)NB or even a legitimate H(e)NB that has been taken to an unauthorized location, the network may incorrectly determines that the H(e)NB is operating in an authorized location. The intention may be to defraud the network operator of revenue by operating the H(e)NB in a region where a particular would have to incur roaming changes, which are typically much higher than that of the charges that are incurred in the user's home region. However, because the H(e)NB is reporting incorrect location, the user deprived himself of the opportunity to receive emergency services should such an emergency arises. Furthermore, the operator may be operating the H(e)NB where it does not have licensed spectrum to operate and may not be able to provide lawful intercept should the law enforcement agency requires it.

Mitigation: Secure location information against the possibility of manipulating location information of a H(e)NB, for example in a trusted environment.

25) Manipulation of external time source

Attacker may try to gain advantage by forcing the H(e)NB's clocks out of synchronization with that of the network. This can happen due to insecure links to external time source or tampering with the procedures that the H(e)NB uses to synchronize its clocks to the networks or may force the H(e)NB into synchronize its clocks to rogue time source. As a result, security procedures that require correct time may fail or produce incorrect result. For example, if an expired certificate is used in authentication and the clock is set to a time before the expiration of the certificate, authentication may succeed. However, if the certificate has been revoked due to expiration, the authentication should not have succeeded. Incorrect time and mis-synchronization can also cause incorrect clock drifts (e.g. clock going too fast or too slow) resulting in other incorrect network operations such as handovers that are timing-based. As a result, the user's quality of service may suffer or handover may fail.

Mitigation: H(e)NB should be notified about information of macro cells from which the H(e)NB can obtain clock information so that it can perform time synchronization based on particular macro cell. A trusted clock server should be located behind the security gateway and communication between the clock server and H(e)NB should have adequate protection. Secure clock synchronization and maintenance functions which are supported in the H(e)NB could be executed within the Trusted Environment.

26) Environmental/side channel attacks against H(e)NB

With physical access to the H(e)NB, the attacker is able to modify some environmental characteristics in the operating the H(e)NB in an attempt to bypass the internal security mechanism or to have the security level of the H(e)NB lowered. This attack poses threats to the H(e)NB's operations and its service lifespan, privacy issues for users and operator network (e.g. exposing network topology).

Mitigation: Provide secure environment and blocks all potential physical access.

27) Attack on OAM and its traffic

OAM system and component of H(e)NBs may be located within the operator's core network or may be located external to the operator's core network where an attacker can attempt to gain access to the OAM and H(e)NB communication link and potentially gaining access to either OAM or H(e)NB. Such attacks may include traffic sniffing attack on the link and man-in-the-middle attack on the link. Since the OAM hosts many important management related functions, such as fault management, configuration parameter management, software managements, etc., such attacks can cause potential denial of service, unauthorized modification of configuration, and disruption of service to the operator's network.

Mitigation: The communication between the H(e)NB and the OAM should be secured.

28) Threat of H(e)NB connectivity to network access

If rogue H(e)NBs are allowed to access to the operator's core network due to the network's inability to verify H(e)NBs profile or other access control information, the rogue H(e)NB may be able to operating in an unauthorized fashion and may provide free service to users that camp on it. Additionally, an attack may also use the gained access to mount further attacks toward the operator's network

Mitigation: H(e)NB SeGW or other network entity in CN should have or can obtain the related profile information, e.g. access control information for H(e)NB, or the status information of the H(e)NB, to check whether a H(e)NB can access the network when it attempts to access the network.

29) Handover to CSG-enabled H(e)NBs

When the users change the allowed CSG list store in the UE, the UE is able to be handed over to a particular H(e)NB with a given CSG ID that the UE does not belong to, even though the handover decision is network-based that takes into account of the CSG list and that the access control to a particular H(e)NB is the responsibility of either the core network or the H(e)NB-GW. If such access control is not performed for any reason, the handover is then allowed and incorrect charges may be incurred by the H(e)NB owner or the operator unable to recoup charges because it is not able to bill

Mitigation: Even on handover the network should check whether the given UE is allowed to access the target H(e)NB.

30) Reporting different H(e)NB ID to SeGW and H(e)NB-GW in Closed Mode Access H(e)NBs

Though the H(e)NB and SeGW are required to mutually authenticate with each other, but once the authentication is successful and an IPsec tunnel is established between H(e)NB and SeGW, the SeGW, based on the current understanding is not obligated to verify any messages encapsulated within the IPsec tunnel. For example, messages between H(e)NB and H(e)NB-GW are integrity and confidentiality protected between H(e)NB and SeGW, SeGW's function is to either peel away or add the protection and forwards them to H(e)NB or H(e)NB-GW without looking at the content. If a modified H(e)NB, after IPsec establishment between itself and SeGW, pretends to be another H(e)NB, a H(e)NB that may be operating in open access mode, the attacker may be able to lure UEs otherwise are not part of the attacker H(e)NB's CSG. The threats to the unsuspecting UEs would be exposure and/or loss of data.

Mitigation: Verify that the identity used by H(e)NB to send control messages toward H(e)NB-GW in the secure tunnel after IPsec establishment is the same identity that is used for establishing the IPsec tunnel.

In the process of studying and analyzing the security threats from every possible angle, the 3GPP security group has defined a set of comprehensive security requirements and security solutions. Though Femtocells network architecture may be somewhat different for Femtocells based on other access technology, the nature of the Femtocells would suggest that the same operating environment and same threats would apply and therefore the same requirements also apply. These requirements are needed to ensure that the Femtocells are operated securely and that every precaution has been accounted for. The technical solutions are also necessary that the requirements are adhered to and allow for cross-vendor interoperability among the different networks and different components of the Femtocells.

3.2.2 Requirements

3.2.2.1 Femtocell Operational Requirements

The secure operation of the Femtocells requires the fulfillment of the operation requirements derived from design stage. These requirements apply to the cryptographic algorithm, algorithm selection, and security parameter configuration and modification. In particular, there are several such requirements:

Req1: Only algorithms of adequate cryptographic strength will be used for authentication and protection of confidentiality and integrity.

Req2: Modifications of Hosting Party controlled information by the operator will only be allowed with the permission of the Hosting Party.

Req3: The extent of Hosting Party controllable information will be controlled by the operator.

Req4: IMSIs of users connected to Femto AP will not be revealed to the Hosting Party of the Femto AP.

The cryptographic algorithm selection requirement comes from the fact that there is a need to use strong algorithms that can resist various crypto-based attacks. As a rule of thumb, all security algorithms for ciphering and integrity protection must have gone through thorough crypto-analysis within the international cryptographic community and that these algorithms are deemed to resist all known attacks within the lifetime of the algorithms. One of the algorithms specified in 3GPP is the Advanced Encryption Standard (i.e. AES algorithm). Published as FIPS-197 in 2001, this algorithm was chosen by the US National Institute of Standards and Technology (NIST) as the de facto encryption algorithm to replace the aging Data Encryption Standard (DES) that had been widely accepted and in use since 1970s. When the DES algorithm was deemed inadequate in the 1990s due to its age, key size and strength, and advances in computing capabilities of hackers and attackers, the NIST called upon international cryptographic community to develop a new algorithm that is stronger and is able to accept bigger key length (DES used 64-bit key as input but only had 56-bit security strength due to the parity checking requirements on the key) as input.

The end result is a competition among noted cryptographer teams that produced the AES algorithm. Some of the notable algorithms that competed to become the AES were Rijndael, Serpent, Twofish, RC6, and MARS. After several rounds of competition, Rijndael came out victorious. It became known as the AES algorithm and has been widely accepted into use in the industry since 2001.

The Hosting Party configuration requirement comes from the fact that the Hosting Party Module, being a commercial-off-the-shelf UICC, is used in ways that is different from the original intended purpose, e.g. in a user equipment or handset. The Hosting Party by definition is a user (usually the subscriber of the Femto service) who has a contractual agreement with the operator to locate the Femto access point within the user's premise and therefore obtaining benefit in the process (e.g. better cellular coverage or broader variety of available services). Since the Femtocell is locally installed, the user may have control over certain parameters to aid the operation of the Femtocell for both the user and the operator. Though the Femtocell is located in the customer's premise, it is still considered a vital piece of network equipment by the operator. To protect other users that connect to the Femtocell, the IMSIs of these users are not revealed to the hosting party.

3.2.2.2 Security Requirements on the Femto Access Point

These requirements are the basic security requirements for the secure operation of the Femto systems and are critical in preventing and countering the threats and attacks identified in the comprehensive analysis stage of the specification building process. They are intended to be comprehensive that they need to be fulfilled completely during design and implementation.

Req5: The integrity of the Femto AP will be validated before any connection into the core network is established.

It is critical that the Femtocell is free from any tampering before its operation starts and before the Femtocell is connected to the operator's core network. This tampering refers to both hardware-based tampering as well as any potential software-based tampering. Since the Femto AP is located at customer's premise, it is impossible to physically safeguard the device itself. If the user is also a skilled attacker, the attacker may try to probe or even modify the device using various tools available to him at the comfort of his home in an attempt to gain unauthorized access, compromise the security and privacy of any potential users that may access the Femto AP or even attempt to obtain free service for financial gains.

Req6: The Femto AP will be authenticated by the SeGW based on a globally unique and permanent Femto AP identity. The authentication will be performed using a certificate provided by the operator, manufacturer or vendor of the Femto AP.

Req7: The Femto AP will authenticate the SeGW. The authentication will take place based on a SeGW certificate.

The FAP has a permanent equipment identity similar to an Electronic Serial Number or ESN in the mobile handset and this identity is not modifiable after the FAP leaves the factory. At time of manufacturing, the FAP may be also pre-provisioned with a device certificate. This device certificate can be issued by the vendor or by the operator if the operator is known at time of manufacturing and the operator has made arrangements to have the certificates delivered to the vendor for installing into the FAPs. The device certificate may also be issued by a third-party trusted by the operator if the operator does not have its own Public Key Infrastructure or is not able to issue certificates. During deployment, the operator may also choose to enroll its certificate in the FAP if the FAP only comes with the vendor certificate. The FAP is also pre-provisioned with the root certificate that is able to validate the certificate used by the Security Gateway both of which are issued by the same CA. Once the device certificate is in place, the FAP and the network will be used to authenticate to the Security Gateway in order for the FAP to gain access to the operator's core network.

Req8: Optionally the hosting party of the Femto AP may be authenticated. The authentication, if used, will be based on EAP-AKA.

If the FAP supports an optional hosting party module (similar to the SIM card used in the handsets), it may also be necessary to perform an authentication of the hosting party using EAP-AKA (Extensible Authentication Protocol) procedures as defined in IETF RFC 4187. An example of the EAP-AKA authentication embedded in an IKEv2 exchange can be found in Chapter 7.

Req9: The Femto AP shall authenticate the FMS, if the FMS is accessed on the public Internet.

Req10: The Femto AP shall be authenticated by the FMS using the same identity as for authentication to the SeGW, if the FMS is accessed on the public Internet.

The Femto Management System helps to manage the operation of the Femtocell and can be located either within the operator's core network or located outside of the operator's network in the public Internet. When the FMS is located within the operator's core network, it is considered secure and therefore no additional authentication is necessary since the FAP would have been authenticated to the security gateway and any traffic between the FAP and any other network elements inside the operator's core network, such as FMS, or AAA server would be considered secure under the hop-by-hop security model. However, if additional security is required, for example, between security gateway and other network elements as some operators may want to implement additional security for various reasons, network domain security can be used between the security gateway and any other network elements located behind the security gateway in the operator's network. Furthermore, the operator may choose, as a policy, to require additional security by requiring the FAP to authenticate the HMS and establish a secure tunnel (using TLS) within the security tunnel that is already in place between the FAP and the security gateway. If, however, the FMS is located outside of the public Internet (e.g. one that is operated by a third party vendor or service provider), this authentication between FAP and FMS is required and that the security tunnel is established as a result of the successful authentication process.

Req11: The configuration and the software of the Femto AP will only be updated in a secure way, i.e. the integrity of the configuration data including the licensed radio parameters and the integrity of the software updates must be verified.

From time to time, configuration parameters and software on the FAP may need to be updated or upgraded, for example when the vendor of the FAP develops updated patches or when the operating environment is changes. Using the secure connection between the FAP and FMS ensures that the delivery of these configuration parameters and/or software update is secure. Furthermore, the content delivered via the secure connection, whether the content contains software modules or configuration parameters, are themselves integrity protected. When the FAP receives the contents, it also needs to verify that the content has not been tampered with or modified in any form before installation to ensure the correct information is being installed.

Req12: Sensitive data including cryptographic keys, authentication credentials, user information, user plane data and control plane data will not be accessible at the Femto AP in plaintext to unauthorized access.

With the location of the FAP being in the customer premise, it is vital that for the FAP to be as "invisible" and "transparent" as possible away from the users. Another word, from the customer's perspective, the customer should be agnostic regarding the fact that there is an actual FAP in the premise. Furthermore, the customer should also be considered unauthorized personnel in terms of accessing data within the FAP, whether the data contains cryptographic keys, authentication credentials, user information, user plane data, control plane data or management plane data. This requirement implies that not only physical security requirements in terms of physical access, but also implies that any remote access capability is strictly enforced to insure only authorized access is allowed. Restricting physical access may be as simple as removing unused ports, such as a HDMI or a USB port and restricting remote access may be as simple as blocking incoming IP traffic that does not originate from either the security gateway or the FMS.

Req13: The time base of the Femto AP will be synchronized to the core network or a trusted network time source.

It is important that the FAP to be time-synchronized to the core network for reasons of operation and security. Since the FAP and the security gateway uses a certificate-based credential for mutual authentication, the certificates expiration date and time must be validated to ensure that neither FAP nor the security gateway are using certificates that are expired, even though the certificates may have a very long expiration date and time due the nature of the FAP. Precise timing is also critical for some Femto systems, such as CDMA-based Femtocells that rely on GPS for better time synchronization.

Req14: The location of the Femto AP will be reliably transferred to the network.

FAP is only allowed to operate within the boundaries of the operator where it has licensed spectrum. The operator relies on accurate location information to ensure that it complies with regulations regarding the operating of licensed spectrum. Accurate location information also helps both the user and the authorities in case of emergency when the authority may need to dispatch emergency service vehicles or personnel to the registered location should such an event calls for. If such an event is life-and-death situation, the result of not getting help to where it is needed due to incorrect location could be devastating.

Req15: Any unauthenticated traffic received from the access network will be filtered out by the Femto AP.

As with any network equipment that is considered to be located on the public Internet, it may be subject to a variety of IP-based packet bombardment from simple PING to more deliberate attacks such as DoS, the FAP should be able to filter out this unauthenticated traffic and not cause any strain to the operator's core network. This also ensures that the FAP has the necessary bandwidth and capacity to do what it is designed to do and not get bogged down processing unauthenticated traffic.

3.2.2.3 Security Requirements on the Security Gateway

The security requirements on the security gateway mirror those on the FAP and both requirements are reciprocating in nature to complement each other, for example, authentication of the Femto Access Point also requires the authentication of the Security Gateway (i.e. mutual authentication).

Req16: The SeGW will be authenticated by the Femto AP using a SeGW certificate. The SeGW certificate will be signed by a CA trusted by the operator.

Req17: The SeGW will authenticate the Femto AP based on Femto AP certificate.

The certificate in the security gateway is issued by the operator or by a CA trusted by the operator. Furthermore, the security gateway also has root certificates to validate the certificate in the FAP. This root certificate is issued by the same CA that issues the certificate for the FAP used in authentication, which can be the vendor CA, operator CA or a CA trusted by the operator. The level of trust in some of the certificate may require that validation path contain up to four certificates.

Req18: The SeGW may authenticate the hosting party of the Femto AP in cooperation with the AAA server using EAP-AKA.

Req19: The SeGW shall allow the Femto AP access to the core network only after successful completion of all required authentications.

If there is a hosting party module present in the FAP and the operator policy dictates that this additional authentication is needed, the security gateway will also perform this authentication procedure with the aid of the AAA server that holds the credentials used in the hosting party module. The hosting party module is similar to the USIM used in UMTS and LTE and relies on a pre-provisioned shared secret between the hosting party module and the AAA server. Since both UMTS and LTE use well-defined AKA procedures (UMTS AKA and EPS-AKA respectively) for authenticating the subscriber to the network using pre-shared credentials, similar procedure is extended using EAP methods (i.e. EAP-AKA) which is also commonly used in other services when 3GPP interworks with other access technology such as WLAN, WiAMX, or CDMA2000 networks. When both the FAP access authentication and the hosting party module authentication, if required, have been successfully completed, the security gateway will grant FAP access to the core network via the security tunnel that is established as a result.

3.2.2.4 Security Requirements on FMS

Req23: The establishment of the secure backhaul link will be based on IKEv2.

Authentication between the FAP and the security gateway using certificates provisioned in the FAP and the security gateway is based on the Internet Key Exchange protocol version 2. This is by far the most popular and secure authentication and key exchange protocol for authentication between two entities running IP protocols. It can accommodate a variety of security credentials, including certificates and pre-shared keys as well as accommodate multiple authentications within one IKEv2 session. The end result of a successful authentication between two entities is the establishment of a secure tunnel between the end points and a set of security associations or (SAs), one for each direction of the security tunnel. A security association is essentially a unidirectional logical connection created for security purposes. All traffic traversing a SA is provided the same security protection. The SA itself is a set of parameters to define security protection between two entities. An IPsec Security Association includes the cryptographic algorithms, the keys, the duration of the keys, and other parameters. The security associations are negotiated to manage the security tunnel in terms of the cipher algorithm to use, traffic encryption keys and other parameters to be passed over this secure connection.   Details of complete IKEv2 protocol are specified in IETF RFC 4306 and are left out for brevity. Interested readers are encouraged to seek out additional information regarding IKEv2 in IETF RFC 4306. The connection between the FAP and the security gateway is also sometimes known as the "backhaul link" into the operator's core network and as a single point into the operator's core network, every security precaution has to be taken into account to ensure the its access, even if the network node is located in the public Internet, as in the case of the FAP.

Req24: The backhaul link will provide integrity protection of the transmitted data. It may provide confidentiality protection of the transmitted data, depending on operator requirements and/or policies.

Req25: The security solution for the backhaul link will be based on IPsec ESP tunnel mode.

Req26: Any connection between the Femto AP and the core network will be tunneled through the Backhaul Link.

One of the reasons and goals of setting up the secure backhaul link is to ensure the security of data between the FAP and the operator's core network. The types of data include control plane data, user plane data and management plane data. Integrity protection over the security tunnel ensures that any data transmitted in the tunnel is free from tampering. However, the operator policy may not require that all tunnel traffic be ciphered. In some deployment scenarios, the backhaul to the operator core network is a PON or GPON network in which case the operator owns the entire network and since the data between FAP and the operator's core network does not go through the public Internet, the operator may choose not to provide additional security (e.g. ciphering) over the backhaul link or may even choose not to deploy security tunnel since the operator may rely on the inherent physical security provided by their underlying PON or GPON network. For most cases, the security of backhaul link is needed and will be based on IPsec ESP tunnel mode. ESP Mode of operation is a mode in which IPsec tunnel is used to protect the whole IP packet and it has the advantage that the entire IP packet including IP packet headers is protected with security transform (e.g. ciphering and/or integrity protection) and re-assembled into another IP packet with its own IP packer header. Another mode of operation supported in IPsec is called transport mode where protection is only applied to the payload of the IP packet, in effect giving protection to higher level layers. This mode of operation, however, is not supported in the Femtocell and in the SeGW.

Req27: The security solution for the backhaul link will be compatible with common network address and port translation variations and support firewall traversal.

To accommodate different broadband deployment scenarios whether they are home-based or enterprise-based, various common network address and port translation schemes are supported. In some enterprise settings, the FAPs are deployed behind a corporate firewall and in this case, firewall traversal is also supported.

With the security requirements firmly defined, specific security mechanisms also put in place to fulfill these requirements and provided the necessary security in the Femtocells.

3.2.3 Femtocell Security Mechanisms

3.2.3.1 Physical Security Mechanism

A truly secure platform starts with a secure environment for which all software, hardware, firmware, and configuration modules are built upon. The secure environment in the Femtocell is called the Trusted Environment (TrE), which allows for secure execution of sensitive functions, such as authentication and encryption, and storage of sensitive data, such as the security credentials. Analogous to the Trusted Environment in the Femtocell is a Trusted Platform Module that is commonly found in today's personal computing platforms, among other general computing devices such as a personal computer or a high end personal computing tablet. This trusted environment defined for the Femtocells differs from that of the user subscriber identity module, commonly known as a USIM. The USIM is built on a removable UICC architecture while the TrE defined for Femtocell is not removable and is part of the overall FAP design.

The Trusted Environment is built on a hardware-based root of trust that is bound to the Femtocell for which the TrE resides and it provides the basis for all other security functions to operate on. A root of trust means that it is not defeatable or simply "as secure as it gets". Even if the root of trust is destroyed, an attacker still cannot gain anything from it like getting the private keys. The TrE starts with a secure boot process every time the Femtocell is turned on or goes through a power cycle. The secure boot process is essentially a device integrity validation that verifies the components or modules before they are loaded. All modules and components have cryptographic checksum that indicate whether the components have been modified. There are several ways in which the device integrity validation can be performed:

1. Autonomous Validation
2. Remote Validation
3. Semi-autonomous Validation
4. Hybrid Validation

In the autonomous validation, the cryptographic checksum are stored locally on the Femtocell. As the Femto is powered up, the system generates a cryptographic checksum for each component or module and compare these values against that of the stored values in the Femtocell. As each component passes the check, the component is then loaded to bring each function within the Femtocell online. For example, the validation may start with the OS related modules, and work its way to the communication related modules and then onto authentication functions and so on so forth. At any point if the validation check is unsuccessful, the validation aborts. The system may attempt to restart and go through the process again or it may initiate some remediation action.

In remote validation, the cryptographic checksum are store remotely in an external entity called Platform Validation Entity or PVE. At the start of the remote validation process, the integrity check is still performed locally by the Femtocell and the results of which are sent to the PVE. The PVE makes a determination from the integrity check results as evidence that the validation is successful and then gives explicit indication to the Femtocell that it can proceed with other procedures to bring up the entire system. It may also be beneficial to consider the PVE to be an entity that sits at the edge of the operator network since one of the purpose of such a validation is to grant or deny access to the network. One benefit of storing the cryptographic checksum of the components and modules externally is to relieve the burden on the FAP's secure storage. Another benefit is that the cryptographic check sum stored in the network will be less prone to tampering.

In semi-autonomous validation, part of the validation is assessed and performed on the Femtocell initially without having any external report or input. This part of the validation only applies to some core components in the Femtocell. If any core component fails the integrity validation check, then the Femtocell should not proceed to start other procedures, such as the authentication procedure with the network. Otherwise, the Femtocell continues with assessment of additional components and engages in an authentication process with the network and signals these results to an external PVE. The PVE can then make further decision and determination to allow the full operation of the Femtocell.

The hybrid-validation is quite similar to that of the semi-autonomous validation but differs in that the validation is performed for different stages of the boot process for initial boot and trusted boot. Validation is performed locally during initial boot, If successful, additional integrity measurements (e.g. for additional modules or components) are collected for both local validation. Both the additional integrity measurements and validation results are sent to the network for further validation and verification. The local policy on the Femtocell and the policy of the operator should determine which measurements should be validated locally and which measurements should be sent to the network validation entity for further validation to help the core network and Femtocell to make access control decisions.

The particular validation approach to take depends on a variety of factors, for example, deployment scenario, implementation, and philosophy of the network operator. Any one of the above approach would satisfy the validation requirement when properly implemented with the necessary components firmly in installed in place. It should be noted that none validation method impacts the interoperation of the Femtocells.

3.2.3.2 Remediation

Remediation is the process to allow the Femtocell to correct validation errors or even runtime errors and may be used anytime the when such an error occurs. Most likely reasons for the Femtocell needing remediation are replacement of modules and components (software) that have become corrupted (due to tampering or due to system glitch). To support remediation, the Femtocell may be placed in some limited access state, for example to allow only a connection to a remediation server. The remediation server (e.g. a FMS that also houses the remediation software modules) then sends over the software/firmware modules to the Femtocell so that the Femtocell can replace the modules in question and try validation or execution again. Figure 3.3 shows how a remediation process might work where the modules in question are sent to the Femtocell one patch at a time after which the Femto AP verifies and installs the patch(es) being downloaded.

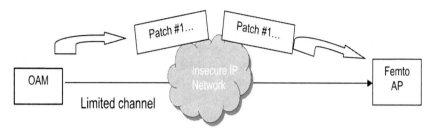

Figure 3.3. Femtocell remediation

After receiving a module (a module could be part of one or more patches as one module may exceed the maximum number of octets allowed in one particular container for a single patch) from the remediation server and before installing the module, the Femtocell performs a validation check similar to the one used during the secure boot process. An example of how this may look like is given in Figure 3.4.

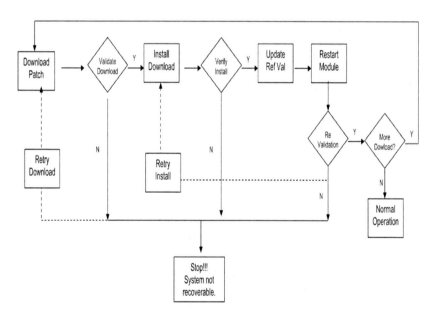

Figure 3.4. Femtocell Remediation Process

If a Femtocell is experiencing severe validation errors, it may be possible that the Femtocell is under attack and the remediation process may need to be done in an incremental fashion where the trust level within the Femtocell has to be built up gradually upon installing the downloaded modules. As the trust levels are built up, more and more functions and components within the Femtocell are validated for secure operations and thus bringing the Femtocell into a secure state for normal operation. This strategy is similar to divide and conquer where smaller victories will lead to bigger and ultimately overall victory.

If none of the above brings the Femtocell to a normal operational state, the Femtocell may attempt recovery by way of a hard reset where the Femtocell is returned to a pristine state with all the factory defaults and configurations before trying the remediation. If this method fails, the last resort is to have service personnel on site to attempt recovery by other means or to have the Femtocell brought into the service center for repair and recovery. It is also possible that a particular implementation would allow a hard-reset of the Femtocell, similar to other consumer devices such as wireless routers. The hard-reset would bring the Femtocell into what is called a "pristine" state – a state that is as if the device itself just came out of a factory. This would then allow the user and the operator to start afresh and trying to rebuild the Femtocell into a normal operational state again.

3.2.3.3 Mutual Authentication and IPsec tunnel establishment

Before the Femtocell is allowed to connect to the operator's core network via the security gateway, it has to be authenticated. This authentication is mutual between the Femtocell and the security gateway. The Femtocell is pre-provisioned with a device certificate. The provisioning of the device certificate can happen at the factory, at point of sale, or at point of deployment. The device certificate is issued by a Certificate Authority (CA). It can be the CA of the operator, the CA of the Femtocell manufacturer, or a third-party CA and all of which must be trusted by the operator. During issuance of the device certificate, the Femtocell manufacturer and the CA have to work together so that an identifier that uses Femtocell identity can be constructed in a Fully Qualified Domain Name (FQDN) format for the device certificate. Further restriction is placed on this FQDN to be not only unique among Femtocells from the same vendor, but also globally unique among Femtocells from all vendors deployed in a particular operator's network. For example, this FQDN can be something like femto123.huawei.com or femto123.nsn.com to identify at least the manufacture. Additional details can be found in Chapter 7 on Security Profiles.

Similarly, the security gateway is also pre-provisioned with a certificate that is issued by a CA trusted by the operator. In some instances, the security gateway may be provisioned with several certificates in order to be able to authenticate Femtocells from different manufacturers should the operator decide not to put restriction on the Femtocells to have certificates issued by the operator CA or by a third-party CA. The identity or identities used in the security gateway certificate or certificates are constructed similar to that of the identities used in the Femtocell device certificates. In case the deployment scenario does not have presence of a DNS server, the FQDN of the security gateway cannot be properly resolved and the security gateway certificate is forced to use the permanently assigned IP address of the security gateway as the identity.

During the mutual authentication between the Femtocell and the security gateway, either Femtocell or the security gateway can check whether the certificate used by the other are revoked for any reason, e.g. expired. The Femtocell may do so using OCSP while the security gateway may do so using either CRLs or OCSP or both. Generally, the CRL or OCSP servers are located in the domains of the manufacturers of the Femtocells and are accessible via the public Internet or located within the operator domain depending on the type of certificates and whose CA ultimately issued the certificates. In either case, the certificates used in authentication would have sufficient information for the revocation checking process to locate the appropriate servers to perform the necessary checking.

Figure 3.5 shows a simple certificate-based mutual authentication using IKEv2 procedures.

A more elaborate and detailed IKEv2 procedure that also includes the optional Hosting Party Module authentication for the 3GPP Femtocell can be found in Chapter 7.

Once the mutual authentication is successful, an IPsec tunnel is established between the Femtocell and the security gateway and this tunnel will be used to encapsulate all traffic between the Femtocell and any other network element inside the operator's network, such as Femto Gateway, MME, etc.

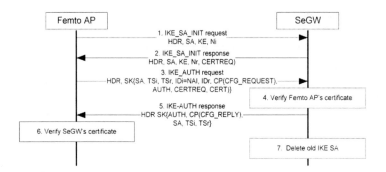

Figure 3.5. Certificate-based mutual authentications between Femto AP
and Security Gateway

Some Femtocells also have a removable Hosting Party Module, which is essentially a Subscriber Identity Module (SIM) that is commonly found in handsets for identifying a subscriber, storing security credentials, phone books, etc. Some operators use this Hosting Party Module for storing configuration parameters or other management parameters that can be used by the Femtocell for connecting to the network, setting up location information, mode of operation, etc. Because of the sensitive nature of many of these parameters, the Hosting Party Module also needs to be authenticated. Since mostly that these modules are off-the-shelf SIM modules, they are pre-provisioned with pre-shared secret that can be used for authentication purpose. The pre-shared secret is shared between the Hosting Party Module and the HSS/HLR of the operator. To support Hosting Party authentication, the Hosting Party Module is needs to support EAP-AKA method for authentication and needs to have an IMSI type of globally unique identity. Since IMSIs are generally used for subscribers, the IMSIs used for the HPM may be a special type of IMSI that is separated from the general IMSI population used for subscribers or for example a range of IMSIs that are dedicated for Femtocells. Additionally, these IMSIs may also be given some attributes specific for Femtocell to avoid the possibility that a user would attempt to use the Hosting Party Module as if it were a regular SIM.   The identification and distinguishing of IMSIs are part of the HSS/HLR process to accommodate Femtocell Hosting Party Module access. Following a successful mutual authentication between Femtocell and the security gateway, the Hosting Party authentication is performed using EAP-AKA with the security gateway acting as an EAP authenticator and forwards the EAP protocol message containing the access request to the

AAA server in order for the AAA to retrieve authentication vector from Authentication Center by way of the HSS/HLR for the underlying AKA authentication procedure. Even though some operator chooses to use Hosting Party Modules for their deployment and management of Femtocells, the Hosting Party authentication is not mandatory and is an operator option. However, if this optional Hosting Party authentication is performed in addition to the mutual authentication between the Femtocell and the security gateway, both must be successful in order to establish the above mentioned IPsec tunnel between the Femtocell and the security gateway in order for the Femtocell to access the operator core network.

3.2.3.4 Location Verification

As stated previously that the importance of location verification cannot be undermined. It is crucial that the operator of the Femtocell services to be able to verify the location of the Femtocell for various security, operational, regulatory, and billing purposes. In the Femtocell deployment where there is a Femto Gateway, the Femto Management System and/or Femto Gateway will be the node that is used to verify the location of the Femtocell. Operators require assurance of the Femtocell location to satisfy various security, regulatory, and operational requirements. This is needed so that the operator will not get into trouble with the law for operating the spectrum license that they spent billions of dollars to acquire outside of area where the operator is allowed to operate these licensed spectrum as the regulatory agencies that oversee these spectrums need to ensure fair use and interference free for other users of adjacent spectrum.

There are many ways that the location information can be verified, depending on implementation, deployment scenarios, and the relation between the wireless network operator and the broadband access provider. The availability, reliability, and accuracy of the location information depend on the method and the information available for which a particular method is applied as well as network configuration and deployment scenarios.

The common methods in which location information can be determined can fall into two categories:

1. Using information provided by the Femtocell
2. Using information provided by the other means

The type of and exact information available from the Femtocell also depends on vendor implementation of the Femtocell, network configuration, and the relationship between the wireless service provider and the broadband service provider that carries the traffic to the operator's core network. When only the public IP address of the broadband access is known and provided by the Femtocell, by binding the physical ports of the broadband access network with the geographical information, the operator can reasonably determine the location of the Femtocell. Many broadband access providers assign their pool of public Internet-routable IP addresses based on geographic locations of the users and therefore it is reasonable to assume that a particular user's IP address, once assigned by the broadband access provider, remains relatively static for a long period of time. In general for this method, the assigned IP address, user identification and location information related to IP address are collected upon initial manual provisioning (or via plug-and-play via SON-type of FMS) and are stored in the network database. The core network component can query the network database to obtain the port number(s) bound with the IP address, and/or the address information (even the longitude and latitude values) when the Femtocell reports its IP address. To support this method, the broadband access provider has to maintain a working relationship with the operator by providing an interface to the wireless service operator operating Femtocell service for it to be able to query the geographic location information based on the IP address. A number of standards have provided such an interface, such as the Network Attachment Subsystem as defined in the TISPAN specifications.

If the IP address is the only information available in determining the location of the Femtocell, there may be some limitation to the reliability, especially when the Femtocell is connected through a local area network or when the Femtocell is connected through a virtual private network. Other additionally available information may aid in increasing the accuracy and reliability of the location information, such as access line location identifier provided by the broadband access provider. In this case, the location identifier is part of the Femtocell profile that is stored in the network's Femtocell home register (similar to the home register that is used to locate a mobile handset).

Some Femtocells are equipped with Global Positional System (GPS) or Assisted-GPS capabilities. These types of Femtocells can provide the most accurate location information based on geo-coordinates. The accuracy sometimes can be within as little as three meters. Because these types of Femtocells require a direct line of sight to an open sky, some operators have guidelines for the users to install the Femtocells near windows of a home. Furthermore, since GPS receivers in the Femtocell need to latch onto and synchronize with GPS satellites via line of sight or without obstruction between them, weather conditions and other environmental variations may impact the ability of the GPS receivers. This is reason why in certain deployment scenarios, the operator recommends in the published user guides that the Femtocell be installed near window so that it can receive adequate and unobstructed signals from GPS satellites. Of course, these types of Femtocells are generally more expensive to produce because of the added hardware to support GPS capabilities.

It is also possible that the mobile network operator has deployed one or more macro base stations near the Femtocell, especially since one of the reasons for the Femtocell is to provide coverage for gaps within the network (e.g. dead spot in the network). To support roaming and handoffs of mobile station between macro base station and Femtocell, the network has to maintain, for each base station including macro base station and Femtocell, a list of neighboring base stations or cells. When a Femtocell is turn on, it scans the neighboring cells and collects the neighbor information such as cell ID, PLMN ID, and other information. If the Femtocell can collect similar information from three neighbors, the network can fairly accurately determine the location of the Femtocell using a method called triangulation. Otherwise, the network has to make best efforts using the available information sent from the Femtocell.

The information available by an UE may also be able to help the operator to determine the location of the Femtocell. As more and more mobile handsets are smart handsets, the capability of these handsets may have increase dramatically in the last few years. Many of which are also equipped with GPS or Assisted-GPS receivers. In these cases, if information provided by a Femtocell is not sufficient or if macro base stations are not available nearby (e.g. rural area), an operator could potentially allow the Femtocell to initially start up and establish a limited connection to the network. The Femtocell is allowed to register UEs that attaches to it. The Femtocell can also query the UE into providing location information (e.g. geo-coordinates) through the UE's internal GPS receivers. Before any service requests are granted, the operator, using the information provided by the UE, makes a determination as to whether the location of the UE and most importantly the location of the Femtocell is valid and authorized. Since the Femtocell has limited range, it is reasonable to assume that the coverage area of the Femtocell that a UE connects to is within the confine of the operator's network coverage area.

For more advanced handsets that support device-to-device applications, it is also foreseeable in the near future that a handset, even though it may not have GPS receiver, can connect to another handset that does have GPS receiver via device-to-device connection and therefore is able to query the location of the other device. Since device-to-device connections are meant to be proximity-based, if the two handsets can made a connection using device-to-device features, it is generally assumed that they are nearby and can offer reasonable confidence to the operator that both are within the coverage of the Femtocell. The operator, in this case, can make an assumption and use the information provided by an UE that does not directly connect to the Femtocell but connects to a UE that is connected to the Femtocell to determine the location of the Femtocell. In many cases, since location determination cannot be made precisely, but best effort and due diligence on the part of the operator may well satisfy the regulatory aspects of the requirement and therefore allow the operator to operate and avoid potential penalties for failing to fulfill regulatory requirements.

3.2.3.5 Access Control

Femtocells are essentially configured in one of three access control mode for their closed subscriber group (CSG):

1. Closed Mode
2. Open Mode
3. Hybrid Mode

In a closed mode CSG, the Femtocells are configured with an IMSI list of mobiles (e.g. UEs) that are allowed to access the Femtocell. The

3 Security of Femtocells

purpose of this is that because the Femtocell is to be located at customer premise, it consumes some resources of the customer such as power and broadband access. Depending on the operator's mode of operation, the Femtocell may be given to the customer, sold to the customer or leased to the customer. Furthermore, the operator may charge the customer a monthly fee for allowing Femtocell to be deployed in the customer premise. When a customer is responsible for supplying the power and broadband access to the Femtocell, the customer may wish to limit the access of the Femtocell to the people whom he or she is acquainted with in an attempt to reduce the resources consumed by the Femtocell, especially in some instances the broadband access is being charged based on metered usage. This is particularly true in some less developed regions where the Femtocells are envisioned to be deployed as an inexpensive way for the operator to quickly extend its network or deploy a new network. In some of these regions, the access to Femtocell and/or broadband may be viewed as a commodity, given the high likelihood that both power supplied to power the Femtocell and the broadband access used to connect the Femtocell to the operator's network are charged at a premium compared to other more developed regions. Other uses of the closed mode CSG are for security reasons, for example to limit the access to only known associates or acquaintances in order to reduce exposure or risk to potential attacks.

In an open mode CSG access, the Femtocells are unrestricted to allow any subscriber to access and establish connection, such as subscribers who are served by the operator or roaming subscribers whose home operator are in a contractual roaming agreement with the serving operator at the Femtocell location. This is envisioned when a Femtocell is considered for a public place similar to that of a Wi-Fi hot-spot, for example in airports, train stations, convention centers, or hotels. Another scenario where this open mode CSG access is useful can also be envisioned as deployment in some less developed regions. For example, in some remote African villages, the entire village may have only one Internet connection and share one Femtocell. Generally, in this case, security is less of a concern and open mode access makes both access and management much simpler when the necessary security precautions have been taken. Other scenario of open mode CSG access can be within certain enterprise deployment, for example, if the deployment location is isolated and it can be ensured that only company personnel or registered guest are within coverage of the Femtocell, there is no strong reason to complicate the configuration by simply leaving the access mode to open.

Hybrid mode CSG access is a combination of both closed and open mode CSG access. In a hybrid mode access Femtocell, the Femtocell may give prioritized access to certain users when capacity is limited. Generally, Femtocells are small cells that have limited capacity to serve the subscribers and if there are too many users accessing the Femtocell simultaneously, some users may not be able to receive service.

When the Femtocell is operating in closed mode access, access control and membership verification of CSG is handled by various network nodes, depending on the capabilities of the UE, the Femtocell, as well as the network deployment scenario. Some Femtocells were rushed to market before the CSG capabilities were included, as are some existing handsets. In this case, if the operator still wishes to enforce access control based on CSG, the access control may be performed by the Femto Gateway. If, however, both the Femtocell and handset support CSG capabilities, the access control is performed by the SGSN/MSC in case of the UMTS type of Femtocells and by the MME in case of the LTE type of Femtocells.

To counter potential threats that a compromised Femtocell may use a legitimate identity (e.g. one that is configured to operate in closed access mode) and credential to connect to the security gateway and establish an IPsec tunnel and then use another identity (e.g. identity of another Femtocell that is configured to be operating in open access mode or hybrid access mode), the Femto Gateway also needs to verify the CSG identity that is sent from by the Femtocell is correctly contained in the UE-related messages. This threat is possible due to the fact that once the IPsec tunnel is established, the Femtocell and the Security Gateway assume the content within the tunnel are secure and that the security gateway's function becomes that of simply stripping out the headers and forwarding the content to the intended network element (e.g. Femto Gateway). If the Femtocell uses another identity for the message to Femto Gateway, the Security Gateway has no way of knowing whether that particular identity is the identity that is used to establish the IPsec tunnel and therefore the threats are possible. The requirements exist in the specification to verify the CSG identity of the UE related messages are indeed pointed to the identity of the Femtocell that is used to establish the IPsec tunnel, but because of the differences in deployment and proprietary implementation, many solutions are possible. One such solution is described below for Femto ID verification.

To have assurance that the FAP ID or mapped identity (collectively referred to as simply FAP ID for the purpose of discussion) is not spoofed by FAP with which it is communicating with the Femto-GW, there need to have a secure way to associate that ID with the secure association that took place at the time the FAP authenticated and attached to the SeGW. If the SeGW and the H(e)NB-GW were integrated together, then the integrated device could have a small database that ties the identities used in Subject Alternative Name fields of the certificates to the FAP ID that is supposed to be assigned to that FAP. The Femto-GW could then filter all of the arriving S1 Setup Request messages (in LTE) to ensure that they do in fact carry that FAP ID.

However, when the SeGW and Femto-GW are physically separated, there needs to be some way to carry the association of the Subject Alternative Name, as given in the FAP device certificate, to the S1 connection being set up to carry control and signaling traffic. One way to do this is to carry a binding between the Subject Alternative Name and the FAP's inner IP address that was assigned by the SeGW to the Femto-GW. Since it is reasonably difficult for any host to spoof an SCTP connection from an IP address that is not routed to it, it is therefore reasonable to assume that if the Femto-GW merely knows the IP address that was assigned to a given FAP, it will be able to prevent other FAPs (which don't have access to the traffic sent to that IP address) to spoof a connection pretending to be the original FAP.

One way to carry the binding is to use the Domain Name System, which already has the capability to support binding names to addresses. The SeGW can update the DNS so that it contains a mapping from the Subject Alternative Name to the inner IP that it assigns to the FAP. It is typical for VPN gateways to perform such a dynamic DNS update when clients connect to them. Then, when the Femto-GW receives an S1 Setup Request message, it can derive a DNS name that corresponds to the FAP ID presented to it, perform a DNS lookup on this DNS name, and verify that the IP address returned is the same one being used as the source of the SCTP connection over which the S1 Setup Request arrived.

Furthermore, the same verification can be extended to other S1 messages originating from the FAP as well. This will provide the verification to satisfy the FAP message verification requirement.

Figure 3.6 depicts this approach for the FAP ID verification process. The terminologies for the network elements and interfaces used in the diagram are commonly found in 3GPP Femtocells.

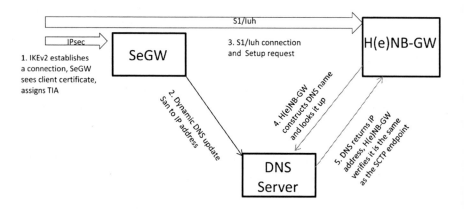

Figure 3.6. DNS-based Femto ID Verification

In this case, the Femto-GW could also contain a database that maps FAP IDs to DNS names, or some other convention for naming the FAPs could be used in the choice of Subject Alternative Names in the certificates; for example, all FAPs can use FAP_ID.MNC.MCC.3gppnetwork.org as the Subject Alternative name field in the certificates. This would make it easier for the Femto-GW to associate an ID with a DNS name that would be used to lookup the inner IP address that was assigned by the SeGW and used to update the DNS server.

The solution described above is simple and easy to implement as most of the necessary components are already in place in most Femtocell deployments.

Access control can also be based on Access Control List (ACL) for initial version of Femtocells that do not support the CSG feature. ACL can be implemented as either a black list which explicitly excludes the members in the list or it can be implemented as a white list in which case, only members in the list are allowed to access. ACL-based access control is performed by the Femtocell because that is the entity where the ACL is hosted. The ACL, either black list or white list can be maintained by the user whose premise is hosting the Femtocell. To manage the list, the user can either use a web interface to connect to the Femtocell ACL access webpage or in some cases use some special short message application (e.g. SMS) to communicate with the Femtocell for such a purpose.

3.2.3.6 FMS Protection

As the number of Femtocells deployed in the network increase, the Femto Management System plays an increasing role in the management and configuration of the overall system operation and maintenance. Essentially, the FMS is composed of a TR-069 manager and a file server. The functions of the FMS include, but not limited to the following:

- Location verification
- Assignment of serving elements such as Serving FMS, Serving Security Gateway or Serving Femto Gateway
- TR-069 Auto Configuration Server (ACS)
- File server for upload and download
- Provisioning of configuration data to the Femtocell
- Performance and Fault updates
- Provides Serving Security Gateway discovery
- Femto software update

TR-069 is a standard for remote management of end-user devices that is commonly deployed in the broadband network. The protocol is used to address increasing number of Internet access devices such as cable modems, DSL modems, routers, gateways, set-up boxes, etc. for the end users. By re-using the standardized protocols and interfaces of TR-069 for the FMS, it greatly enhances the flexibility and compatibilities of the FMS and at the same time greatly reduces the initial development cost. Essentially, the FMS can be an off-the-shelf TR-069 manager with minimum enhancement for Femto duties and because of the off-the-shelf nature and some operator's desire to outsource the management of the

Femto devices to third-party service provider or management entities, it is possible for the operator to deploy a FMS either within the operator's network or outside of the operator's network (i.e. in the open Internet). Access to the FMS and traffic (e.g. management plane) between the FMS and Femtocell should be enforced with necessary protection to allow either case.

For the FMS that is located outside of the operator's network in the public Internet, it is exposed to the openness of the Internet and may be constantly subject to the attacks. Access to the FMS shall be applied on after a successful TLS authentication and subsequent tunnel establishment. The mutual authentication between the FMS and the Femtocell shall be based on pre-provisioned device certificate in the Femtocell and a network certificate in FMS. This device certificate in the Femtocell may be provisioned by the vendor, the operator or a third-party trusted by the operator. Note that a device certificate is also used for device authentication. The certificate used for device authentication may also be the same certificate that is used for FMS authentication or it may be different, depending on the operator policy in which certificates are being used. Both the Femtocell and the FMS may choose an array of options for certificate validation for example using CRL method or accessing an OCSP for checking the revocation status. The accessibility of the FMS in the public Internet makes both CRL or OCSP checking of the certificates easier to manage because the Femtocell already has access to the Internet and is not subject to accessing only an operator core network entity before authentication to the operator core network is established.

When the FMS is located in the operator's core network, the access to the FMS and the traffic between the FMS is protected using one of two options, depending on operator's policy:

- FMS traffic is protected hop-by-hop, with the first hop being between the Femtocell and the Security Gateway and the next hop being between the security gateway and the FMS. Since the Femtocell needs to establish IPsec tunnel to the security gateway in order to gain access to the network, it is a given that any traffic that originates from the Femtocell to the network, whether it is user plane, control plane or management plane traffic (i.e. FMS traffic), must go through the IPsec tunnel. There is no additional procedure to be performed. Since the next hop between the security gateway and the FMS is well within the domain of the operator's network, the operator may simply no apply additional security because the operator may consider everything within its domain secure and therefore conclude that no additional security is necessary or the operator may choose to apply additional security. If the operator believes that the path between the FMS and the security gateway is not secure, it may choose to apply additional security using network domain security mechanisms that are commonly used between network elements in the 3GPP network.

• FMS traffic is protected end-to-end between the Femtocell and the FMS using TLS based authentication and tunnel establishment procedures similar to that of when the FMS is located in the public Internet. Note that the FMS traffic in this end-to-end case is first encapsulated in and IPsec tunnel between the Femtocell and the security gateway and since this IPsec tunnel between already exists for other traffic such as control plane and user plane traffic, this FMS traffic inside the IPsec tunnel is transparent to the security gateway and is untouched by the security gateway before forwarding to the FMS.

3.2.3.7 Clock Protection and Synchronization

Clock is an important aspect of the Femtocell. Correct time and synchronization are needed for security reasons as well as operational reasons that require different degrees and granularity of precision. For example, certificate revocation and expiration checking may not require the same time synchronization precision as exchange of signaling messages. Though the Femto is equipped with a clock, but its clock may have become mis-synchronized due to clock drift, spike in power, or any other reasons that may have caused the mis-synchronization. In order to support time synchronization functions with the time server, which may reside either within the operator's network or may reside in the network as a network clock server? If the clock server is located inside the operator's network, the IPsec tunnel between the Femto and the security gateway can be used to protect the initial hop of traffic, similar to that of the FMS protection. When the clock server is not inside the operator's core network, the communication between the Femtocell and the clock server has to be secured. Two such commonly used clock synchronization protocols are Precision Time Protocol (PTP) and Network Time Protocol (NTP).

For verification of certificate expiration or revocation before the establishment of the IPsec tunnel to the network, the FAP uses a time value that it saved in its secure memory when the FAP was last powered off. Though the internal clock in the Femtocell would continue to advance once the power to the Femto is restored, it would cause the clock to become out of synchronization (even in case that the internal clock may have a separate backup source of power). But because the Femtocell may not have access to a real-time clock to synchronize its clock before validating the certificate, it would have to be synchronized to a secured time server afterwards once the connection to the core network is established. Furthermore, the certificate validation may also need to be performed again once if the synchronized local clock differs from the secured time source significantly, enough to warrant such a rechecking of certificate expiration and/or revocation.

3.2.4 Security of UE Mobility in Femtocells

UE mobility is the process of the UE moving from one base station (e.g. macro base station) to another base station (e.g. macro base station or a Femtocell) or vice versa while maintaining security context among other critical parameters so that the UE can be served in another cell immediately or maintain the active session that it had. Two types of mobility events are possible for an UE, namely idle mode mobility and active handoff. In a normal macro to macro mobility event, the Closed Subscriber Group membership does not come into play and if capacity, link condition, or other condition allows, the UE is moved from one base station (e.g. source) to another base station (e.g. target). However, if the target base station is a Femtocell operating in either closed access mode or hybrid access mode, access control or membership verification is required to be performed on the UE. The UE can be either capable of supporting Closed Subscriber Group feature or not capable of supporting the Closed Subscriber Group feature and in either case, access control is still required. The access control or membership verification is in addition to any normal mobility procedures, such as security context information transfer, etc. and takes precedence over any other normal mobility procedures. Another word, if the target Femtocell is not allowed to serve the incoming UE, the UE cannot be acquired by the target Femtocell and therefore cannot be served by the target Femtocell. In this case the UE will still be served by the source until radio condition no longer allows the connection to be maintained and the UE will be dropped out of coverage. This particular case also applies if the source base station is also a Femtocell. From mobility point of view, it makes no difference to the mobility procedure whether the source base station is macro base station or another Femtocell.

When the source is a Femtocell and the target is a macro base station, the normal mobility procedures are followed since the macro base station does not perform Closed Subscriber Group membership verification for incoming UEs.

To support UE mobility events better among Femtocells, it is desirable to have direct interfaces among the Femtocells, much like the direct interfaces between the macro base stations. An advantage of direct interfaces is faster mobility for the UEs since control plane signaling needed may be exchanged between two adjacent Femtocells directly without going through the security gateway and the core network. Additionally, user plane data may also be carried on the direct link realizing the same performance gain as the control plane signaling. One issue arising from the use of direct link between two Femtocells is that the Femtocells need to authenticate each other. This can be done by using certificate-based credentials or shared-secret-based credentials. However, if certificate is used, what and whose certificate to use needs to be determined and agreed up in advance, especially since the two Femtocells may be manufactured by two different vendors. Since the Femtocells are equipped with vendor certificate when the equipment is initially provisioned into service in an operator's network, Femtocells manufactured by two different manufacturers will not be able to authenticate with each other using the vendor certificate. It is presumable that two vendors' certificates are verifiable only to the root certificate of the vendor, unless the two vendors utilize the same Certificate Authority to issue their certificates. One solution is for an operator to require their vendors to install every other vendor's root certificate in the Femtocell, but this would no doubt create an issue since vendors are also competitors and are unlikely to agree to such an arrangement. Another potential solution is to install pre-shared key in all the Femtocells. As the number of Femtocells grows, the number of direct links one Femtocell has to maintain with other Femtocells also grows makes the use of pre-shared key impractical. Since security requires that each direct link must use different pre-shared key to secure, the number of pre-shared key that need to be installed and maintained becomes a nightmare. There are those who argue that since the Femtocells are CSG-membership based, it is unlikely that one Femtocell would need to maintain too many direct links as the UE's subscriber owner is not expected to perform mobility events too frequent with too many of his members. However, this argument fades quickly in an enterprise setting. Another solution is using the certificate enrollment procedures defined for the macro base station case. Since macro base station comes out of the vendor's factory the same way as the Femtocell (i.e. with vendor certificate pre-installed), the operator usually requires that an operator issued certificate to replace the certificate issued by the vendor. CMPv2-based certificate enrollment procedures in RFC 6712 have been defined in 3GPP TS 33.310 for macro base stations to replace the vendor certificates. This procedure requires the CA or RA to be equipped with root certificates of all vendors that the operator uses, which is easier done than requiring vendor base stations to install root certificates of all other vendors. Once all the base stations in the operator's network have been enrolled with the operator issued the certificate, authentication between base stations using these certificate is much simpler and has no interoperability issues. Femtocells supporting direct interface features and thus mobility features can also be enrolled in such a way as the macro base stations. One major difference in using this enrollment process between macro base stations and Femtocells is that the enrolment procedure may take place with the RA/CA accessible on the public Internet where as the enrolment procedure for macro base stations must always go through the operator's core network. In case the enrolment goes through the security gateway, the security gateway must allow the vendor certificate for this procedure and later also allow the vendor certificate for backhaul establishment, even though at the completion of the enrolment procedure, the Femtocell would have acquired an operator issued certificate whereas in the case of the macro base station, the vendor device certificate is no longer allowed to be used to connect to the operator's backhaul.

3.2.5 Femto Security and LIPA

LIPA, or Local IP Access, is possible if the Femtocell supports such a feature. LIPA allows IP-capable UEs to access the Internet locally without having the Internet traffic going thought the operator's core network elements such as a Serving Gateway or a SGSN via a network component called the Local Gateway. The Local Gateway can be either co-located with the Femtocell or located externally to the Femtocell but connected to the Femtocell via direct physical interfaces. Routing Internet traffic to the UE through the Local Gateway is an operator option that is controlled by the operator as to when Internet traffic is allowed through the Local Gateway or not. There are cases, for example, for supporting lawful intercept, where the operator would prefer the traffic to go through the Serving Gateway or even the SGSN. Additionally, the operator would always maintain control plane signaling via the core network so that it can use its charging functions, depending charging model for Local Gateway routed Internet traffic. Since the Local Gateway is also considered as an operator-owned but user premise deployed equipment, it presents another potential possibility for attackers to enter the Femtocell system to take advantage of the threats previously identified. However, securing the Local Gateway is done using the requirements similar to that of the Femtocell. If the Local Gateway is not co-located, the interfaces to the Femtocell should also be secured, for example, via secure tunnel between the Local Gateway and the Femtocell using either pre-shared key or other means. Local Gateway may use the same IPsec tunnel when it connects to other network elements in the operator network as the Femtocell via the security gateway. Furthermore, the security gateway may allocate the same IP address as the Femtocell or it may allocate a different IP address to the Local Gateway traffic depending on operator policy. Generally, the Femtocell has ways of differentiating the IP address allocated to itself or to the Local Gateway, since the IPsec tunnel end point for the Local Gateway also terminates in the Femtocell in the co-located case. The Femtocell can use information sent in the IKEv2 CFG payload or sent directly in the IPsec tunnel via DHCP or OAM protocols to distinguish the two different IP addresses.

3.2.6 Emergency Services Support in Femtocells

Emergency services in the Femto systems are supported in the Femtocell, Femto Gateway, or both. The only difference in terms of handling emergency services when compared to that of the macro base station case is that the Femto system would not perform access control or CSG membership verification when an UE attached to the Femtocell is attempting to initiate emergency services. However, the system should also ensure that a compromised Femtocell manipulating control signaling for initiating emergency call to make a non-emergency call in fooling the system in believing that the control signaling is indeed for emergency is not allowed to take place. For example, the network may perform additional digit analysis on the emergency call initiating messaging to verify that an emergency number has been dialed before bypassing CSG verification. This verification should also be applied after the emergency call ended so ensure non-emergency calls would not be allowed to be piggybacked, for example, as part of the control signaling. Additionally, location verification also plays a role in supporting emergency service as the emergency service center may need the correct location of the Femtocell in order to correctly dispatch the necessary personnel to provide the emergency service requested by the user in case the user is unable to provide it to the service center for any reason.

3.2.7 Femtocell Security Profiles

Security profiles provide details as to how a particular security protocol is used when it is being used in an environment that was not considered in the original design, or how a particular algorithm is used within a particular protocol. These profiles include the use of certificates in the Femtocells, the handling of certificates, as well as the use of IKEv2 for authentication and IPsec tunnel establishment. Details of these security profiles can aid further understanding of the security of the Femtocells. Details can be found in a later Chapter 7.

3.3 Case Study on Attacks of the Femtocells

Ever since the first commercial deployment of Femto system, there had been reports of hacks and attacks on various Femtocells, some are severe and damaging while others are only cause the operator to lose revenue. These attacks had been widely reported in various blogs or forums. While the publicity can be damaging to the reputation of the operator as well as the industry, but it nevertheless raise awareness in terms of security and standards compliance as all of these attacks and hacks of the Femtocells were made in what the industry calls non-standards compliant systems.

3.3.1 Case 1

In one operator's network, the earlier versions of the Femtocell did not perform location verification. The operator was giving Femtocells to its customers in certain regions after receiving complaint of poor signals. These customers did not pay a monthly subscription fee for using the Femtocells. However, in other areas of the same operator network, the operator was charging the customer a marginal monthly fee. Some users learned about the monthly fee waiver and called customer service to change their billing address to that of one where the customers' monthly fees are waived. Since the Femtocells do not perform location verification, the users were not charged with the monthly subscription fee for using the Femtocell.

3.3.2 Case 2

The Femtocells in another operator's network was cracked open by attacker to review its internals: couple of system-on-chips (SoC), radio chip, GPS module and other hardware, including some tamper-detection hardware. After sometime, the hacker was able to physically tap into the SoCs and able to recover data dump as the Femtocell was rebooted. Hacker's tools were used to recover the SSH and ROOT account passwords, and in a coincidence out of neglect the vendor also happened to have implemented the same password. This practically allowed full access the Femtocell from the inside and all kinds of attacks then became possible.

3.3.3 Case 3

The attacks on the Femtocells in the of another operator's network that cost as little as £50 were made known in July of 2011 by a group known as The Hacker's Choice (THC). The attack involves some reverse engineering of the Femtocells and seemed simply enough. Though the serial port of the FAP had been sealed off as the designers followed the correct requirements to prevent physical access, the attackers were able to figure out the pin connection and made some ingenious soldering to connect to the serial console of the Femtocell. Once inside the console, the attacker was able to guess the root password that is used for administrator access by looking at reversed engineered system firmware. After retrieving the root password and gaining access as a root user, the attackers were able to gain further details regarding the configuration parameters as well as details about the IPsec tunnel between the access point and the security gateway. From that point, the attack has total control of the Femtocell and was able to disable the firewall, change the internal configurations and other settings. Once the configurations had been modified to the hacker's desire, the privacy of any user that was accessing his or her wireless services through the Femtocell was at the mercy of the attacker. The operator whose Femtocells that were subject to these attacks acknowledged that these attacks had been possible on the earlier versions of the Femtocells that were deployed in 2009 and prior but had since upgraded the older versions of the Femtocells that were used in these attacks and patched the security hole before the attack became publicly available.

3.3.4 Case 4

In yet another attack on yet another operator's Femtocells as demonstrated in a Black Hat 2011 event in August of 2011 and subsequently published in other forum, the researcher was able to take advantage of the Femto system's recovery procedure by connecting to an unauthenticated server instead of an authenticated Femto Management Server, install his own software, firmware image and configurations, and turn the Femto into his playground. The recovery process is normally used by the Femtocell as a way to reset to factory setting should malfunction occurs (e.g. due to software incorrectly loaded at boot time or at execution time). When rebooting does not resolve the software related issues, the next recourse to take is using the recovery procedure built into the product (also known as remediation process) to reset the Femtocell to factory pristine condition and then attempt to connect to an authenticated TR-069-type of server for the reinstall any software or patch that have been upgraded since the device left the factory.

Sometimes, the recovery procedure built into the product is also sometimes called a "limp-home" mode. In the version of the Femtocell being attacked, the firmware's server is not authenticated and along with the public key in the parameter list that is not digitally signed, and the Femtocell is able to accept anyone's firmware module without verification, including that of the attacker's firmware image. Of course, the researcher that performed this attack has lot of help at his disposal and knows the Femto system and security inside out. The description of the may be overly simplifying, but the attack was real. As a result, the attacker was able mount a number of attacks and compromising both system security and user privacy. Though the attack was focused on some of the older version of the Femtocells, the holes have since been plugged but the threats and security issues are real.

3.3.5 Case 5

In the most recent reported attack in July of 2013 and demonstrated in the Black Hat 2013 event in August of 2013. In this demonstration, an earlier version of a Femtocell that was deployed in an US operator's network was used as an example. The HDMI port on the bottom of the device was not disabled when the device left the factory. Inside the Femtocell is a compact version of the Linux operating system, providing basic functions as well as other essential system functions. Through the HDMI port, the researchers were able to communicate with the Femtocell and subsequently gain root access to the Linux system. With the root access to the system, the researchers implemented methods to intercept the calls and SMS messaging that went through the Femtocell. Apparently, this older version of the Femtocell that was vulnerable did not have updated version of the system software and did not properly implement a trusted environment. The operator in this case also acknowledged this vulnerability and provided software upgrade to the affected models.

In all of the above mentioned attacks, only certain versions of particular non-standards compliant Femtocells were used to perform the published attacks – the versions of the Femtocells that were manufactured prior to standards publications or specifications, for example specifications produced by 3GPP or 3GPP2 such. In general, these pre-standards Femtocells were built as a way to reach the market quickly and did not follow the standards and were not based on requirements and solutions that were part of the published standards. They were built before standards existed or when the standards were still in development stages. These standards producing organizations contributed countless hours to create the standards, in the process provided thorough security analysis, threats analysis, and careful security solutions design to ensure that the Femtocells can resist potential threats and attacks discovered in the course of standards building. While it is difficult to grasp the vendor development thought process when the initial pre-standards Femtocells were build, it would have also been prudent to follow good industry practices and security best practices, for example, in choosing passwords carefully. The possibility that some of the Femtocells attacked may not have followed industry best practices or security best practices during the development process may have in fact contributed to the ease of which passwords were guessed in some of the above mention attacks. Note that all of these attacks were performed by security researchers and no actual customer data had been compromised as a result of these attacks to the best knowledge of the author. At the present, the networks of all operators, including the ones whose Femtocells experienced one form of hacking or another remain secure.

4

CDMA Femtocells

Femtocells are not exclusively for 3GPP (e.g. UMTS and LTE). In fact, as early as in the year 2007 and before anyone of the European or Asian operators deployed any Femtocell service, US-based operator Sprint (Sprint has since been acquired by a Japanese operator Softbank in July 2013) had already introduced Femtocell service to its subscribers in a product called Airave. The Sprint Femtocells, like versions introduced by yet another competing US operator Verizon, are based on CDMA technology, or Code Division Multiple Access. This technology was developed in the early 1990s as a second generation digital wireless access technology to replace the aging analog access technology. The earliest version of the technology was published in a series of specifications by TR45 (Mobile and Personal Communication Systems Standards) sub-working group within TIA (Telecommunication Industry Association), the standards developing organization responsible for developing telecommunication standards in the U.S. In late 1990s, the standards developing organization in various parts of the world joint forces to form what is known as 3GPP2 or the 3rd Generation Partnership Project 2. 3GPP2 is collaboration between members of various telecommunication associations to make a globally applicable third generation (3G) mobile phone system specification within the scope of the ITU's IMT-2000 project. In practice, 3GPP2 is the standardization group for CDMA2000, the set of 3G standards based on the earlier CDMA technology for 2G (known as IS-95 as published by TIA in the U.S.). This earlier version of the CDMA standards has been superseded by IS-2000 (i.e. CDMA2000 in 3GPP2 terminologies) as the CDMA technology went from 2G to 3G in the early 2000s. The 3G CDMA technology based on the CDMA2000 specifications is capable of supporting up to three carriers but most commonly supporting only one carrier, commonly known as CDMA 3G1X. In addition to circuit-switched voice services supported in CDMA networks, packet switched data, 1xEVDO, and IMS services are all supported. Since 2009, the evolution of CDMA-based technology and evolution of WCDMA-based technology have merged in what is currently known as LTE, or Long Term Evolution as a 4G technology.

Several CDMA Femtocell variants are supported depending on the underlying CDMA architecture. This support also depends very much on the deployment scenarios. In particular, the following Femtocell architecture variants are supported, at least published in the standards for CDMA2000:

- IOS-based Femtocells with MSC (Figure 4.1)
- IOS-based Femtocells with MSCe (Figure 4.2)
- IMS-based Femtocells (Figure 4.3)
- HRPD/CDMA 1x packet Femtocells (Figure 4.4)

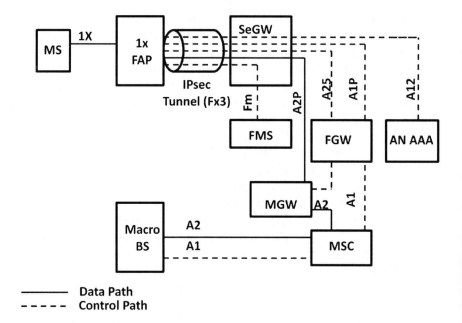

Figure 4.1. CDMA Femto IOS-based 1x Voice Architecture with MSC

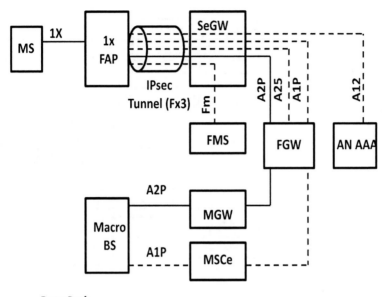

Data Path
----- Control Path

Figure 4.2. CDMA Femto IOS-based 1x Voice Architecture with MSCe

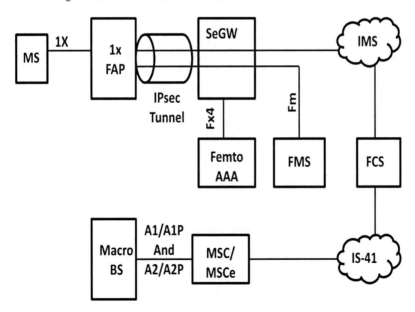

Figure 4.3. cdma2000 SIP-based 1x CS Service Femto Architecture

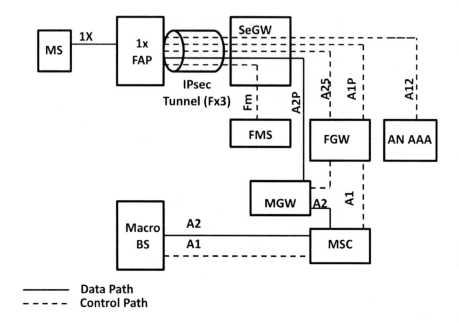

Figure 4.4. HRPD/cdma2000 1x Packet Femto Architecture

Although there is the slight overall architecture variation in the CDMA
Femtocells intended purpose of whether the architecture is IOS-based or
SIP based and whether the architecture is intended to support 1x Voice,
1x CS, or 1x Packet services, the security architecture remains strikingly
similar to the 3GPP version.   IOS-based simply means that the
specifications for the various interfaces are specified in a series of 3GPP
specifications known as the Interoperability Specifications for CDMA
2000 Access Network Interfaces while SIP-based means that the
interfaces are based on Session Initial Protocol as specified in IETF RFC
3261.

The following network elements are considered essential part in the
overall CDMA2000 Femtocell architecture:

- Femtocell Access Point (FAP): is the home base station in the
  3GPP2 network. It is the equivalent of the Home NodeB or Home
  eNodeB in the 3GPP networks. It is also known as a wireless access
  point or base station that operates in licensed spectrum to connect a

3GPP2 capable mobile station to the operator's network through the public Internet infrastructure

- Femtocell Authentication, Authorization and Accounting Server (Femto AAA): provides Femtocell Access Point authorization, access control policies, and billing functions (depending on the billing model of the operator). This server may be a standalone AAA server or may be a separate function within an AAA or an Access Network AAA in the operator's network, if capacity allows for such a dual purpose AAA server.

- Femto Gateway (FGW): is part of the operator's core network that provides aggregation, proxy, and signal routing functions. Though it is critical for the overall operation of the Femtocell services, but it is not essential in terms of providing security related function in the security architecture reference model.

- Femtocell Legacy Convergence Server (FLCS): the equivalent of the Femto Gateway used in certain deployment a scenario (e.g. SIP-based) that is used to provide aggregation, proxy, and signal routing functions for the FAPs to access services within the system in the operator's core network.

- Femto Management System (FMS): a network entity that resides in an operator's network and aids in auto-configuration of the FAP before the mobile station can access services through the FAP. It can also be used as a server to support TR-069 type of functions for software and firmware download. In addition, FMS also supports configuration for access control in Femtocell systems. FMS can also reside outside the operator's network in the public Internet, for example, if the operator chooses to let a third-party provider to manage the Femtocells, among other operational reasons or arrangements.

- Foreign Agent/Mobile Access Gateway (FA/MAG): FA is used to de-tunnel and deliver datagram to mobile nodes that were tunneled by the mobile node's home agent while the MAG is a function on an access router that manages the mobility-related signaling for a mobile node that is attached to its access link. These entities are needed when mobile IP is supported in the FAP.

- Security Gateway (Security Gateway): provides a single point of entry into the operator's core network from the Internet for the FAP to securely access the operator's core network. SeGW is also responsible for providing authentication and authorization of the FAP to establish IPsec tunnel between the SeGW and the FAP.

- Femtocell Convergence Server (FCS): is part of the support for SIP-based Femto service in CDMA 2000 networks that provides an equivalent functions to an MSC/VLR in the macro networks, for example in providing processing and control for calls and services. It is an IMS Application Server that interworks FAP that supports cdma2000 1x mobiles, SIP environment of IMS and the appropriate MAP network elements in the CDMA 2000 networks. Some of the MAP network elements are HLR, MC, MPC, MSC, etc.

In the context of Femtocell architecture, some of these network elements may not be specific to the Femtocell and may serve other purposes as well, as in the case of the AAA server. It is likely that a particular operator's deployment will have only one AAA server which may serve dual purpose of both being a Femto AAA and an access network AAA when packet data services are supported.

4.1 Variations in CDMA Femto Architecture

From the various architecture figures presented in the previous section, it can be shown that there are namely two major variants of the CDMA Femtocell architecture, one that is based on SIP signaling and one that is based on IOS signaling. The major difference between these two variants is in the operator's core network behind the security gateway. For the Femtocell architecture based on the SIP-based signaling, the Fx1 interface extends from the FAP to Media Gateway (MGW) and the Fx2 interface extends from the FAP to the IMS core network. Fx1 is the interface that carries the RTP packets or Real-time Transport Protocol payload format that has been defined for use in 3GPP2 network while Fx2 is the interface that carries the SIP signaling packets. Since IMS is all-IP, the IP-based protocols for both user plane and control plane are used in the IMS-based Femtocell Architecture. Although the MGW is not shown explicitly in the simplified SIP-based Femtocell Architecture, it is well understood that this network element is part of the IMS core network. Furthermore, there is no MSC/VLR within the IMS core network, which requires a conversion function or interworking function. Another difference between the SIP-based variant and the IOS-based variant is that there is no Femtocell Gateway in the SIP-based architecture. Instead, a Femtocell Legacy Convergence Server (FLCS) or a Femtocell Convergence Server (FCS) is used that provides functionality in the IMS core network that mirrors the MSC/VLR in the traditional macro network. Since the FAP is identical irrespective of the underlying core network, whether it is IMS or 3GPP2 network, to interwork with IMS, the FAP and FCS cannot communicate using the legacy interface (A1 or A1P and A2 or A2P) between the BS and the MSC. From the perspective of a macro BS, the Femtocell Convergence Server looks exactly like another MSC. For interface purposes, the FCS fully supports the necessary interfaces and the associated protocols for inter-MSC communications and operations.

Since 3GPP2 also supports packet data services through either CDMA2000 1x or CDMA2000 HRPD air-interfaces, there may be cases where the FAPs are both CDMA2000 1x- and CDMA2000 HRPD-capable. In the Femtocell Architecture supporting CDMA2000 HRPD-capable FAPs, common entities are re-used as much as possible. It is also necessary to extend the user plane interface and control plane interface from the FAP to the packet data serving node or PDSN. PDSN can be considered the equivalent of PGW within the 3GPP network. In the architecture supporting either CDMA2000 1x or HRPD or both, the Femto Gateway provides the aggregation function for various interfaces as well as proxy functions for the FAP to access variety of data services in the operator's networks. There are several of these interfaces that the FGW is capable of aggregating, namely:

- A1P: carries signaling information between FAP and the MSC
- A10: carries user traffic between the PCF and the PDSN
- A11: carries signaling information between the PCF and the PDSN
- A12: carries signaling information related to access/terminal authentication between the Access Node and the Access Node-AAA

- A13: carries signaling information between the source AN and the target AN to support inter-AN mobility events such as dormant or active handoffs
- A16: carries signaling information between the source AN and the target AN for HRPD Inter-AN mobility events
- A24: carries buffered user data from the source AN to the target AN for an MS that is in the process of going through an active handoff
- A25: carries user traffic between the FAP and the MGW

In a nutshell, in the IOS-based architecture, FGW is used while in the SIP-based architecture the FCS/FLCS is used, along with the difference in the interfaces as defined for either of the core supported.

4.2 CDMA Femtocell Security

In security architecture of the CDMA Femtocell looks almost identical to that of the security architecture in the 3GPP networks, given the slight differences in terminologies used in the different networks.

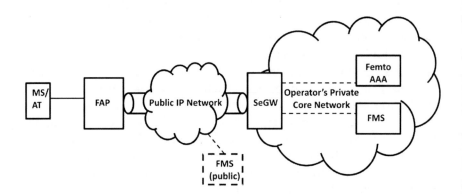

Figure 4.5. CDMA Femtocell Security Architecture

As in the 3GPP2-based Femtocell, the mobile station uses the legacy cdmaOne (e.g. IS-95) and CDMA2000 air interface to access Femto services via the FAP. The FAP connects to the operator's private core network through the public IP network via a security gateway. Since the public IP network is assumed to be unsecure, access to the operator's core network is through the security gateway is authenticated. Successful mutual authentication between the FAP and the SeGW also nets a secure tunnel that is used to carry all user data and signaling data between the FAP and the operator's network. A Femto Gateway (Not shown in the CDMA Femtocell Security Architecture Reference Model) that is part of the operator's core network provides aggregation, proxy, and signal routing functions for user data and signaling, since each FAP is capable of supporting multiple simultaneous MS/ATs accessing services in the operator's network. Though the FGW is critical for the overall operation of the Femto services, but it is not essential in terms of providing security related function in the security architecture reference model and therefore it is not shown for architectural simplicity. Similar FCS/FLCS is also not shown in the CDMA Femtocell security architecture reference model that is needed to support SIP-based Femto services. FMS, which can be located either within the operator's network or outside of the operator's network, is the management server that can be used for provisioning purposes. Current specification defines the FMS as implementing the full protocol of TR-069 as defined by the Broadband Forum's "CPE WAN Management Protocol". In addition to being used to configure and monitor the operation of FAP, the FMS is also used as a tool for the operator to install software updates, firmware updates or perform other management functions. The Femto AAA holds the functions of FAP authorization, access control policies, or potential billing.

4.2.1 FAP Security Features and Mechanisms

The FAP, being in the center of the Femtocell services in 3GPP2 networks, hosts a list of complete security features. These security features, part of inseparable overall features of the CDMA Femto Access Point, are necessary for the secure operation of the Femto services in 3GPP2 networks. These security features include FAP identity, authentication, authorization, security services (integrity protection and confidentiality protection), and FMS protection. These security features, along with the security mechanisms and procedures that are designed for the Femto services to operate in security harmony, define the overall security of the 3GPP2 Femto services.

4.2.1.1 FAP Identity

The CDMA FAP is uniquely identified by an equipment identifier that is similar to that of MEID for handsets. The MEID or Mobile Equipment Identification is the manufacturer's electronic identification of a particular piece of equipment, namely the handset. It consists of a 56-bit number along with a 2-bit check code that has the following format:

- 8-bit region code
- 24-bit manufacturer's code
- 24-bit manufacturer assigned serial number
- 2-bit check code

The globally unique MEID is used for variety of purposes such as authentication and device validation. FAP Identity, also known as FEID, takes on similar format as the MEID meaning that it also consists of a vendor identity and a serial number that is unique within the vendor, making the whole FAP identity globally unique. Though the serial number component of the FEID is local within the FAP manufacturer, however, when it is combined with the manufacturer's identity, it becomes unique globally. Furthermore, the FEIDs are provided by the manufacturer to the operator at time of delivery of the equipment and it is assumed that these identifiers remain constant throughout the life of the FAP equipments. Should the manufacturer run out of identifiers for the FAP equipment, the manufacturer may choose to reclaim the serial numbers of the equipment that have either become obsolete or are no longer being used within the system. However, any reclaiming schemes must be taken with caution as these identifiers are sometimes indexed uniquely in many operational databases in a particular operator's network for a variety of purposes, for example billing. One wonders why the user's subscription number (i.e. cell phone number) is not used for some of these purposes. The reality is that both the subscription number and mobile equipment numbers are used, especially in some of the legacy systems where the only way to identify the piece of equipment is by using the MEID. Furthermore, as the users may choose to sell, to trade or to exchange their equipment without changing their subscription, the combination of both subscription and the MEDI becomes even more useful in identifying both the user and the equipment the user is using. Another reason for the operator to be able to identify the equipment is for operational reasons such as to recall certain equipment known to have defects. A lesson learned operating the

CDMA or 3GPP2 networks is that MEIDs tend to run out quickly and the reclaiming process may not be that efficient. The FEID takes on larger size to counter the potential that they may run out as quickly as the MEID should the Femtocells becomes as popular as the wireless services themselves. As a result, the FEID is also compliant with an extended 64-bit format as defined by IEEE's numbering group. An advantage of the IEEE's extended format is that it is compatible with the traditional 48-bit MAC address format as well as a new 64-bit EUI (Extended Unique Identifier) format.

As the FAP is delivered to the operator, the operator enters the identity in its database, along with binding of the FAP's security credentials, whether it is a pre-shared secret or a device certificate. The security credentials of the FAP may be provisioned at factory, at which case, the combination of the FAP identity along with the matching credential will be sent to the operator, most likely in a batch format. In case of device certificate, for the manufacturers that do not have a CA or the capability of generating their own certificates, the process is a little complicated because the certificates have to come from other source, a source that the operator also trusts, before they are sent to the manufacture if they are to be pre-installed at the factory or if the operator chooses to install the certificates, then the operators would take on the provisioning role for the FAPs and their device certificates.

4.2.1.2 FAP Secure Environment

Similar to the FAP defined in 3GPP, the CDMA version of the FAP are built based on a trusted environment or a secure environment. These so-called trusted or environments are increasingly finding their ways into both operator's equipments as well as user equipment. Traditionally considered secure by means of physical security (e.g. located in secure sites or physically challenging sites such as roof-top or high poles), base stations have increasingly shrunken in size as the FAPs have found their ways into customer's premise. The increasingly exposed nodes such as the FAP are also increasingly relying on trusted or secure environments to compensate for their lack of physical secure due to being deployed in exposed locations such as at a customer's premise or in public locations. Besides providing physical security that is capable of detecting and deterring tampering, the other purposes of the secure environment is to provide a basis of secure storage of critical parameters such as the private keys (e.g. for pre-shared key based schemes or private keys used in the certificate based schemes) and secure operation of many critical security functions, such as cryptographic ciphering/deciphering operations, authentication functions and validation functions so that the keys and security functions are not exposed to the hackers and attackers, even when the FAP is compromised. The secure environment, may be either logical entity implemented via secure hardware or a physical entity, is separately implemented entity inside the FAP. As with the 3GPP secure environment, the basic characteristics of the CDMA FAP secure environment is at start up, the FAP secure environment performs integrity validation of every component within the secure environment before loading or starting those components. The process of building up the components and modules within the secure environment is that as at before these components are loaded in the secure storage within the secure environment, a reference value or a cryptographic check sum of each component is computed and verified. The component or module, along with the verified reference value is installed. Part of the integrity validation of every component is calculating the reference value of the component that is stored in secure storage, comparing the calculated reference value of the component to that of the one stored in secure storage, verifying that they are the same before determining that the component is verified before loading the component into secure execution environment. In addition to the components within the secure environment, the FAP may also perform validation to other components, such as other non-core software modules, configuration modules, or communication modules that are outside of the secure environment before bringing the entire FAP system online after all the critical and essential components have been verified and loaded into secure execution environment.

4.2.2 Authentication

Identical to the UMTS and LTE version of the FAP, the CDMA FAP needs to be authenticated before connecting to the security gateway for accessing the operator's core network. Both the FAP and security gateway are provisoned with certificates that are used as the basis for performing authentication with the security gateway. The FAP device certificate may be issued by the FAP manufacturer, the operator, or a third party trusted by the operator. The security gateway is usually provisioned with the operator's certificate. Since both the FAP and the security gateway need to mutually authenticate each other using the certificate provisioned in them respectively, it may also be necessary for the FAP or the security gateway to provision a trusted root certificate. In case the security gateway is provisioned with the operator's certificate, it may become the trusted root certificate that also needs to be installed in the FAP, regardless of what device certificate is installed in the FAP, whether is the vendor certificate or a third-party CA certifiate. In case the trusted root certificate and the device certificate are not issued by the same entity, the authentication of the FAP and the network would lead to validating each other's certificate, most likely through a chain of trust that has a path leading to the trusted root certificate store in the FAP and the security gateway respectively. The 3GPP2 Femtocell does not define the number in the certificate chain that leads to the trusted root as in the 3GPP case. In the 3GPP Femtocell, that number is four (4). The reason to limit that number to four is that anything above four may lead to authentication failure due to delay induced time out in some of the certificate validation schemes as the number in the certificate validation chain grows. Further more, longer delay in authentication may also lead to performance related issues that may in adverse cause other failures.

4.2.3 Authorization

In the CDMA Femtoell, the FAP also goes through an explicit authorization stage during the initial setup to allow the home network operator to authorize the FAP to continue service. This authorization process is analagous to a "Femtocell service subscription" similar to that of a user's cellular service subscription. This subscription-type of serivce model is to allow the operator to accomodate a variety of billing models and therefore a variety of services. It has been widely reported in some online fora and blogs that when a customer calls in the operator's customer service center to complain about spotty services within the customer's residence, the operator has the discretion to offer a Femtocell to the customer for free or even for very little cost as a good will gesture to retain the customer. Wireless service is such a competitive market place and the operator is willing to go extra miles to keep the customers happy. Sometimes, the operator may even waive or reduce the montly subscription fee associated with offering Femtocell services for these customers with spotty coverage. Generally, in the US, the several operators offering Femtocell services have charged their customers anywhere from $99 to $200 for the FAPs in addition to a monthly service fee of $5 to $10.

In the FAP authorization process, the operator verifies the FEID as provided by the FAP during successful device authentication against the FAP profile information stored in the operator's database (e.g. FAP AAA). Such profile information may include the FAP identity, the FAP's subscription information, CSG information, the service area that is FAP is allowed to operate in, or other billing related information. In case FAP is allowed to operate in other service areas, the FAP may still need to go back to the original serving security gateway in order to go through device authentication if the local serving security gateway cannot handle FAPs that are typically outside of the serving area. This is unlike user authentication in typical wireless service where the user is authenticated by the HLR. Although not stated explicitly, as part of the authorization, it is assumed that the FAP is also checked against the location where it is allowed to operate, using commonly known methods such as IP address being reported by the FAP, GPS coordinates reported by the FAP, or even approximate location as determined and reported by neighboring macro base station to determine the location of the FAP. Since cdma-based FAP requires much higher precision in terms of time synchronization for its operation, there are general requirements during the installation and placement of the FAP itself that it is to be placed near windows or other locations where if direct line of sight is not possible then the obstacle between the FAP and the GPS satellites must be as little as possible. Becasue of the time synchronization requirements, it is very likely that many, if not most of these CDMA based FAPs are equipped with GPS receivers from the factory and can correctly report their locations with much higher precision than any of the IP-addressed based or macro base station reporting methods can possibly afford. In general, the location as being reported by the CDMA FAPs are more reliable becasue of of pontential GPS support and the ability of GPS receivers to provide more accurate location information.

4.2.4 Integrity and Confidentiality Protection

As with any basic security services, the first hop into the operator's core network goes through the public Internet that ends at the security gateway requires the communication link to be both integrity- and confidentiality-protected. This is easily done when the FAP and the SeGW sets up IPsec tunnel after initial device authentication of the FAP between these two entities. The difference with 3GPP Femtocell is that in the 3GPP Femtocell, the IPsec tunnel is optional to set up after the device authentication to accomodate some deployment scenarios where the operator also owns and controls the initial access network (e.g. PON or GPON network) that leads to the security gateway and the operator is confident enough that its security gateway is not accessible from the public Internet, sometimes the operator chooses not to require additional protection to conserve some network resources. In the case of the cdma Femtocell, it is required that IPsec be always set up after device authentication. However, whether confidentiality protection is to be applied shall be at the discretion of the operator based on operator policy. This is due to some regulation if certain regions or countries that ciphering is prohibited. The IPsec usage (based on 3GPP Femtocells) in Femtocell case is further profiled in Chapter 7 in more detail as certain usages of the IPsec would not apply in the Femtocell case. As with any other protocols, though the entire protocol may have been incorporated into the Femtocell during the development stage and taken from readily available off-the-shelf implementations without modification, it is very likely that some of the features of the off-the-shelf implementation of the protocol itself are not very useful or do not apply.

4.2.5 FMS Security

Depending on deployment scenario, FMS may be inside the operator's network in which case there is a security gateway at the edge of the operator's network or the FMS may be outside of the opertor's network in which case a FAP can connect directly to it. There may be some cases where the operator thinks it would be advantageous not be manage the FMS by relying on a third-party management entity to manage it on the operator's behalf for a variety of reasons. For example, since the FMS also implements CPE WAN management protocols as defined in Broadband Forum's TR-069 specification along with the associated functionalities, it may make sense for the operator to let the broadband operator to manage the FMS. Another example is that when the FAP may be shared among two or more operators in what is called a "RAN-sharing" scenario where the FAP may need to host multiple identities and/or security credentials for each of the operator that shares this node, it may make sense for a third-party that is dedicated or expertised in the management of the FAP to take on the management role.

Regardless of the deployment scenario, when the FMS is inside the operator's network, the link between FMS and the security gateway may be protected by operator's internal network domain security mechanism (e.g. IPsec), the security mechanism that the operator has in place to protect the interfaces between internal network elements while the link between the FAP and the security gateway is protected by the IPsec established as a result of device authethentication of the FAP. This is known as a case of hop-by-hop protection of the FMS link, relying on the transitive trust among the FAP, the security gateway and the internal network element within the operator's network. In addition on top of whether this network domain security is applied, the FAP and the FMS may also be protected via a TLS tunnel. However, when the FAP is outside of the operator's network, the FAP and FMS link will be protected by TLS tunnel. In either case, all traffic between FAP and FMS are considered to be integrity and confidentiality protected, by either end-to-end protection via TLS tunnel or by hop-by-hop fashion via IPsec tunnel from FAP to security gateway and network domain security from security gateway and FMS. Whether the FMS link is protected via hop-by-hop or end-to-end, the security is certificate-based. It is possible that the FAP is provisioned with more than one device certificates, for example, one certificate is used for device authentication and setting up IPsec tunnel with the security gateway and yet another certificate is used for mutual authentication with the FMS and establishing secure tunnel with the FMS (e.g. using TLS tunnel).

Since FMS is responsible for providing FAP with configuration parameters and software update functions, such transfer is integrity protected provided by TR-069 protocols and security or by other means, such as hop-by-hop or end-to-end protection described above. FMS, implementing TR-069 type of database maintains hundres of different parameters related to the FAP and the operations of the FAP. Some examples of these parameters include software versions, hardware versions, security-related parameters such as keys or passwords, parameters about neighboring macro cells or FAPs, alarms regarding error conditions, etc. Addition examples of some of these (security-related) parameters can be found in Appendix A. For a complete details of such parameters, please refer to 3GPP2 specification or TR-069 specifications.

4.3 Security Mechanisms and Procedures

The security mechanisms and procedures in the cdma Femtocell include the following:

- Device integrity validation
- Device authentication
- Authorization
- Integrity Protection
- Confidentiality Protection
- FAP-FMS Security
- Signed File Transfer

These mechanisms and procedures very much mirrors similar mechanisms and procedures that have been described in greater details in Chapter 3 for the 3GPP Femtocells and therefore are omitted here. Below highlights some of the differences in these mechanisms and procedures, especially for authorization and signed file transfer that are not described in detail in the previous sections.

After a successful device integrity validation of the FAP and before the first MS can be served by the FAP, the FAP needs to be complete the neighborhood discovery process, establish network connectivity so that it can properly discover and locate the security gateway in order to establish secure tunnel. Thought this is essentially not considered part of the security mechanism or procedure, it is nevertheless an essential process that the FAP must go through in order to proceed with the next process. The parameters that are needed for the neighborhood discovery process (e.g. Home Domain Name, SeGW's FQDN or IP address, etc.) may only need to be set up once and may be done so manually at deployment stage, via remote access, or pre-configuration at point of sale or at the operator office, or may be configured dynamically via the help of the FMS. Depending on deployment scenarios, some operators may deploy a provisioning security gateway to establish the FAP's first connecting point and use the security gateway as part of the provisioning process to further locate the service security gateway, FMS or any other entities such as the FGW. In this case, only the initial provisioning security gateway's essential information (e.g. IP address) for the FAP to discover where it is may need to be installed into the FAP or via DHCP or DNS servers.

After the successful mutual authentication and as part of the establishment of a secure tunnel between the FAP and the security gateway, the security gateway also checks the FAP profile for further authorization. The FAP profile may contain information about the user's subscription, the network operator policy, location authorization, and other relevant information. This profile information is usually stored in the Femtocell AAA and accessed by the security gateway via RADIUS or DIAMETER protocol in the form of AAA access request exchanges. This authorization can be based on for a number of factors, such as whether the customer has paid its bills for using femto service, and all of these factors that are considered for authorization would be detailed in the operator's policy for granting such an authorization. Once both authentication and authorization are successful, it then concludes the IKEv2 procedure by establishing IPsec tunnel between the FAP and the security gateway so that the rest of the remaining initialization procedures may continue, such as location determination and configuration download. Note that only when IPsec tunnel has been established, the communication to the core network is then considered secure with both the integrity and confidentiality of the IP packets sent through the IPsec tunnel between the FAP and security gateway protected using IPsec ESP as defined in IETF RFC 4303. As also mentioned previously, this authorization is also analagous to part of the location verification process as described in the 3GPP Femtocell case.

Even though the FMS and the FAP are protected securely via either hop-by-hop (i.e. FAP-SeGW and SeGW-FMS) or end-to-end (i.e. FAP-FMS), part of the TR-069 also defines a security mechanism called Signed Package Format that can be used to further provide integrity protection for the content of the file transfer between the FMS and the FAP, whether is file transfer is for the provisioning parameters or for any software or firmware download. This file transfer operation is signed cryptographically using the private key that correspond to the device certificate of the FAP. Note that the device certificate of the FAP may be a vendor certificate that is used for FAP to SeGW authentication or another certificate that is used exclusively between the FAP and the

FMS for both FAP to FMS mutual authentication and secure tunnel establishment. In order for this mechanism to work, the signature field of the Signed Package Format contains at least one signature signed by an entity trusted by the FAP along with a certificate or a certificate chain that can lead to a path where the FAP can verify this signature in the signed package. Upon receiving the file in Signed Package Format, the FAP then verifies both the certificate and the signature. If both of these verification succeeds, the FAP will take the received file as being valid and proceeds with installation, execution, etc., otherwise, the downloaded file needs to be discarded and error condition reported to the FMS for further action. Further action may follow the remediation process as described in Chapter 3.

4.4 Differences between CDMA and UMTS/LTE Femtocells

Many of the differences between the CDMA Femtocells and the 3GPP Femtocells have been described in one way or another when the specific security features, security mechanisms and security procedures are discussed in details in the previous sections. Additional details and summary are provided below to capture the differences.

4.4.1 SIM Card Support

In terms of securtiy, one of the major differences between 3GPP Femto to the CDMA femto is the use of optional Hosting Party Module. Though the Hosting Party Module in the 3GPP Femto is optional, but it can be used by the operator for configuration of parameter specific to the operator's environment, much in the way of the SIM card in that is used in the handset of GSM systems. This use of the optional Hosting Party Module in 3GPP simplifies deployment greatly in many cases. Using the optional HPM also affords the option of the operator to perform additional authentication and/or authorization based on the shared secret shared between the HPM and the HLR in the operator's network. The use of SIM card in the CDMA Femtocells is explicitly not supported as the Femto Access Point is considered as a network equipment and not an end user equipment.

4.4.2 Use of Pre-shared Key

Another one of the major differences is in which some of the FAPs manufacturered prior to 3GPP2 Femto standardization may use pre-shared secrets between the FAP and the core network elements instead of certificates as authentication credentials. In the 3GPP Femtocells (manufactured to be compatible with UEs prior to 3GPP Release 8), the FAP may be equipped with HPM or simply called USIM (the HPM prior to Release 8 was simply called the USIM), and AKA procedure identical to that used between the UE and the AuC is used for the authentication of the FAP where as in the cdma case, even though shared secret was used, it was not provisioned via SIM cards. This difference was contributed to the fact that SIM cards were widely used in GSM and UMTS as the defacto standard for distributing user security credentials while the SIM card (the equivalent was called the removable user identity module) was mostly absent, especially in established CDMA markets around the world, with the major exception being the China market. The major operator of CDMA2000 service in China at the time (i.e. China Unicom) when IS-95 standards were being developed had lots of experience operating GSM networks and wanted a way of distributing the user credentials much in the way of GSM credentials were distributed. As a result of the Chinese operator's push and the vendors who were willing to accomodate the regional requirements of the Chinese CDMA operator, the outcome was the removable user identify modules (R-UIM)option that was initially introduced and subsequently adopted in the 3G version of the CDMA (e.g. CDMA2000) specifications. Another reason why the Chinese operator wanted to re-use the SIM cards for cdma was that in the China market, the SIM cards were marketed by the operator as another form of currency for the mobile services as in the early days of GSM and CDMA services in China, the wireless services were marketed as a pre-paid service in which the users need to pre-purchase or pre-load currency into their pre-paid account. Since the SIM card was used to identity the user and also for which money can be added to the account, the SIM card in China became a form of currency equivalent of money for the pre-paid wireless service, even though the SIM cards are relatively inexpensive to manufacture even at that time. The mobile network operator would only recognize the SIM cards and not the users who owned the SIM cards for a very long time during the initial proliferation of wireless services in China.

4.4.3 Use of Device Certificate

For the certificates used in authentication, there are some differences as well. The Femtocells have device certificates that are configured (i.e. provisioned) during manufacturing process. The certificates are signed by a Certificate Authority trusted by the operator. In the 3GPP case, the operators can use certificate enrollment procedures defined in TS 33.310 to replace the device certificate with a certificate that is issued by the operator's CA. This has the benefit for the users and operators in that it simplifies the root certificate installation in that the device and the security gateway now only needs the root certificate of the operator instead of the root certificate of every vendor. Furthermore, in the 3GPP case, this also helps the Femto-to-Femto handover case because now the operator issued certificate in the Femtocells can be used to establish Femto-to-Femto connections (e.g. X2 connection in the 3GPP LTE Femtos and Iurh for the 3GPP UMTS Femtos) for faster handover of the mobile connection. Without these operator issued certificates, UE handovers would have to traverse the Femto-SeGW-Femto path. Of course, provisioning of operator issued certificates require that the certificate enrollment infrastructure to be present and that the Femtocells support the required procedures.

4.4.4 FAP Authorization

CDMA FAP goes through a FAP authorization procedure after the initial device authentication and before the IPsec tunnel is established. While the 3GPP Femtocells do not go through such an explicit authorization process, but some of the information regarding the subscription of the FAP, such as location information is verified also in an explicit location verification process.

4.4.5 Signed File Transfer

While 3GPP Femtocells do not make any assumptions on the underlying protocols which are used to transfer information between the FMS and the FAP, it relies on the mechanism defined to provide the necessary protection. Whether there are additional mechanisms offered by the underlying protocol or not, 3GPP Femtocells are completely agnostic about it. Another word, 3GPP Femtocells consider the security mechanisms, such as the Signed Packet Format in TR-069 as being transparent. This is true since the use of such security mechanism is outside the scope of 3GPP in which case 3GPP cannot mandate it. Rather, 3GPP relies on the security mechanism it defines for its products.

4.4.6 Optional Use of IPsec

IKEv2 is used as the authentication and key exchange protocol for setting up the IPsec tunnel and Security Association between the FAP and Security Gateway. However, in the 3GPP networks, the use of IPsec tunnel between the FAP and the Security Gateway is optional. There are some network deployments where the operator owns both the broadband network as well as the wireless access network. When the operator owns the broadband networks (e.g. GPON), the operator can make certain assumptions of the security offered by the broadband network, which is considered as a private network as it is not of the public Internet. For example, the operator can consider that the security from the broadband access point in the customer premise all the way to the operator's wireless core network secure enough so that IPsec usage becomes unnecessarily burdensome in terms of performance for its customers and for the network. The operator does not see the benefit of the IPsec as it is being used in the traditional Femto systems as the path from the FAP to the operator's core does not traverse the public Internet. Even in this case, IKEv2 usage for device authentication is still mandatory and the operator has to provide some sort of guarantee that the IPsec security would be provided by the equivalent of layer 2 security that is offered by the underlying broadband network which the operator also owns and operates. In the CDMA Femtocells, the use of IKEv2 and IPsec is always mandatory and the case where the operator owns both the broadband access network as well as the wireless access network is not considered.

4.4.7 Direct Interface between FAPs

In the 3GPP Femto, the direct interface is being used in some Femtocells to aid faster handoffs between FAPs, especially in enterprise scenarios and in large residence deployment scenarios where UE handoffs are considered frequent. The support of the direct interface requires the device certificates of the FAPs, which may be manufacture red by different vendors, to be enrolled in operator's CA so that the FAPs can easily authenticate each other and setting up the IPsec tunnel. Though this direct interface feature may be easily supported in CDMA Femto as well, as of now there is no information about its current or future support. No enhancements are expected.

4.4.8 CSG Handling

Because UE handoffs are supported between FAPs in 3GPP Femtocells, before a target FAP can receive a UE during the handoff, the closed access mode FAP has to verify that the UE is also in its Closed Subscriber Group before allowing the handoff to complete. If the target FAP cannot verify that the UE being handed off is in its CSG, the target cannot allow the handoff to complete and the handoff would fail. In this case, the UE/MS must be dropped from coverage. The CSG handling is identical to that of a macro based station to FAP handoff. Since CDMA Femto FAP-to-FAP UE handoff is not explicitly covered, this is yet another point of difference between these two types of Femtocells.

4.4.9 FAP Identity Verification

In the 3GPP Femtocells, the threat that a compromised FAP may modifies its access mode upon establishment of the IPsec tunnel between FAP and SeGW is considered serious enough that additional requirements are put into place to explicitly counter this. The assumption is that after the IPsec tunnel is established, the content of the FAP IPsec traffic will no longer be verified at the security gateway and that the security gateway will simply forward the message onto its next destination based on the IP address routing information in the IPsec headers, for example a message that is destined to the FGW. Since FGW assumes that the contents of the IPsec tunnel is secure, it would normally not perform any additional verification and take the content as being legitimate. However, if the FAP is capable of modifying the content of the messages after the IPsec tunnel is set up, it may change its access mode from closed to open and send such a request to the network. Since the network believes that the content sent in the IPsec tunnel is secure, it also believes that the request to change the access mode of the FAP from close to open is legitimate and grants such a request. However, because the FAP is compromised, by turning the closed access mode to open, user communication and privacy may be exposed. This is just one example of how this threat can cause harm. To counter this threat, 3GPP put in place a requirement that even after the IPsec tunnel is established, there must be verification on the network side that the identity used to create the message sent in the IPsec tunnel is the one and the same as the one that the entity originally set up the IPsec tunnel. This requirement only applies when the access mode in the FAP is set to closed mode access and is exclusive in the 3GPP Femtocells. This threat was either not considered in the CDMA Femtocell designs or is considered unrealistic. In either case, the solutions, one particular possibility was previously discussed in more detains in Chapter 3, are at the discretion of the implementers and the operators who choose to use them as the solutions have no impact on the interoperability of the overall system of the Femtocells. Other solutions are also possible in addressing this threat. Below is a non exhaustive list of other such potential solutions:

- An interface between SeGW and FGW, using an already standard protocol
- Authentication function in FGW
- Authentication function in other network entities (e.g., HLR in case of USIM based authentication)
- Using an already standard protocol between FGW and an entity (e.g., DHCP Server) that holds the allocated IPsec address

Any one of the above mentioned solution may be implemented in such a way to satisfy the requirement of Femto identity verification put forth by the standards and are not in conflict with interoperation of the overall system.

4.4.10 UE or MS Authentication

As with any other CDMA2000 base station, the CDMA2000 FAP also uses a mechanism called global challenge (see Annex A.3) where a random number is generated and broadcasted by the serving base station (i.e. the FAP) so that mobile stations listening on the broadcast channels of the base station can very quickly pre-calculate an authentication response based on the broadcast random number. This global challenge is used to support MS prior to a particular of the CDMA2000 specification and allows the mobile stations to establish authentication and key agreement with the network quickly. In case that the FAP decides not to accept the pre-established response based on the broadcast number, the FAP may also send a unique random number specifically that that authentication instance. This process is called the unique challenge in CDMA2000. Both the global challenge and unique challenge procedures are not used in 3GPP-based systems (i.e. UMTS and LTE). For newer releases of the MS in CDMA2000 networks, the FAP also supports AKA, the same authentication and key agreement that is used in UMTS.

4.4.11 LIPA Access

In CDMA2000 Femtocells, LIPA access is only applicable when the FAP supports the HRPD air interface. As the LIPA is basically an IP access bypassing the operator's data core network, it would only make sense if the fundamental services are based on IP, as in the case of HRPD in the CDMA Femto and in 3GPP Femto. In the CDMA Femto that supports only voice, LIPA is not supported at all.

4.4.12 Security Profiles

Other differences and variations in which the certificates, IPsec, TLS, and IKEv2 are used between CDMA Femtocells do not affect the operation of these Femtocells. For example, profile of IKEv2 relies more on the specification in IETF RFC 4945 and 5280 and places fewer restrictions on the certificates about the use of outdated algorithms (e.g. SHA-1 support). In IPsec, it is also possible for the CDMA Femtocells to negotiate confidentiality protection with transform IDs for encryption ENCR_NULL. Another example is the mandatory to use and recommended to use public key size s for Diffie-Hellman are 1024 and 2048 respectively, which is less stringent on the counter part used and recommended in 3GPP Femtocells.

4.4.13 Other Differences

There are also other differences between the CDMA Femtocells and the 3GPP Femtocells that are attributed to the differences in fundamental designs and operations of these two distinct access technologies.

Interested readers are encouraged to further explore those differences in greater details via other avenues.

5

WiMAX Femtocells

The WiMAX access technology was considered as an early 4G technology that was designed to compete with LTE in 3GPP and 3GPP2. It is based on air interface of the IEEE's 802.16 family of standards for wireless broadband access. The goal of WiMAX was to provide data rate of anywhere from 30 Mbits/second all the way up to 1Gbits/second for wireless high speed data access. It is viewed as both a mobile wireless service supporting full mobility as well as a potential replacement fixed broadband access (i.e. the last mile access as sometimes is called) due to its higher data rate that may be comparable to that of the traditional fixed broadband access. Since its introduction, WiMAX-based wireless access systems have been gradually deployed in many countries and many more markets around the world due to its relatively low cost of deployment compared to other access technologies such as CDMA and GSM. This is because when compared to traditional CDMA or GSM system design, the macro base stations of the WiMAX system tend to provide a coverage area that is much larger, making it easier deployment targets in large metropolitan areas or other areas with dense population. Another attractive reason for some of the popularities in operator choosing WiMAX is that its deployment does not require any backward compatibility to be supported as it is a fresh deployment. While data speed for many users may be its strong suit, extended coverage far lags behind other traditional and established traditional cellular networks.

5.1 WiMAX Architecture and the Femto

Similar to CDMA and GSM, WiMAX architecture is divided into access network and core network. The access part of the network in WiMAX is called the ASN, or Access Service Network which consists of a base station connected to an Access Service Network Gateway or ASN-GW. The core part of the network in WiMAX is called the Connectivity

Service Network or CSN that consists of AAA and HA, or Home Agent. The WiMAX architecture is designed in a way that anyone with the necessary spectrum can set up an access network and become a network access provider or NAP connects to a network service provider or NSP that provides the necessary IP-based backhaul connecting to the Internet. The NSP is considered as the Internet service provider within the WiMAX reference model.

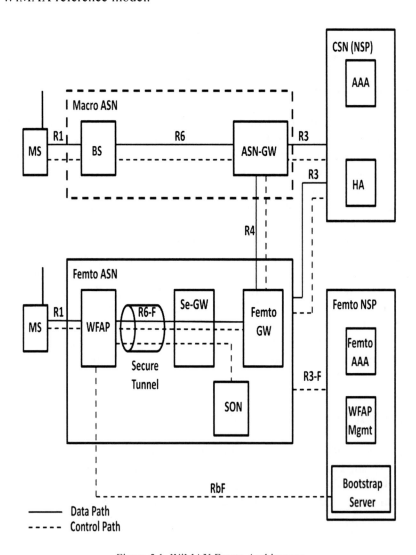

Figure 5.1. WiMAX Femto Architecture

Figure 5.1 shows a Femtocells architecture based on the WiMAX architecture. The WiMAX architecture is also clearly shown as part of this overall architecture and how WiMAX and WiMAX Femto interwork with each other.

As with other wireless access technologies, though the terminologies are of slight variation, but the functional components are strikingly similar. The devices that provide access to WiMAX are called Subscriber Stations as opposed to Mobile Stations in CDMA and UE in 3GPP. The Femto base station is called the WFAP in WiMAX. The management server in WiMAX Femto is called the WFMS. In addition, there is a SON, or Self Organizing Network server that provides additional management functions for plug-and-play deployment and provisioning. The details of these functional components in the WiMAX Femto are the following and their essentially mirror those in the 3GPP and/or CDMA systems:

5.2 WiMAX Femto Functional Components

Functionally, the WiMAX Femto is composed of the following network elements:

- WiMAX Femto Access Point (WFA)
- Security Gateway (Se-GW)
- Self Organizing Network Server (SON)
- WiMAX Femto Management System (WFMS)
- WiMAX Bootstrap Server (WBS)

5.2.1 Femto Access Points (WFAP)

WFAP not only implements the entire base functions of the WiMAX macro BS in a small package, but also provides additional functionalities. It provides user access connectivity over the R1RP interface. The R1RP interface Provide Access Connectivity to a user over R1RP interface is the equivalent of the R1 interface in the macro WiMAX case where it connects the user's MS to the BS. Since access control is an important aspect of Femto services, it is also a function that is implemented and managed by the WFAP. The WFAP maintains and manages a Closed Subscriber Group list to allow or disallow (e.g. white list or black list) a

MS to access the WiMAX network through the WFAP. The users have options of using various input methods to update the CSG since the WFAP is expected to be located at customer premise and the customer who provides power, Internet connectivity and real estate most likely has the role of creating, updating, and/or deleting members in the CSG that is to be hosted in the WFAP. Of course, the updating of the CSG may also be triggered at the network side by the management server, for example, if a particular MS's subscription has been terminated, the management server may inform the WFAP to remove the MS from its CSG. The WFAP also supports the control of radio resources, dictating whether and when to turn on and off the air interface of the WFAP and possibly other maneuvers to prevent neighboring WFAPs to interfere with each other. The SON server also plays an important part controlling the radio resources as SON is expected to have a full map of the network. Knowing where the WFAPs and allocating different frequencies for those WFAPs that may interfere with each other would help tremendously in terms of spectral efficiency gain and user satisfaction. As part of the radio resource management, WiMAX defines both Radio Resource Controller (RRC) RRC and Radio Resource Agent (RRA) functions for supporting radio resource management (RRM) which are functional elements and signaling to transfer the RRM context between the BS and the ASN-GW (and another BS). If RRM is to be supported by the WFAP, the WFAP is also expected to implement both the RRC and RRA functions. The WFAP also connects to the Bootstrap Server, WFAP Management System, Femto Gateway and Security Gateway for a variety of operational and security procedures. Since MS is also expected to be handed off between WFAP and macro BS to support full MS mobility, the WFAP also supports mobility functions. Part of the mobility support in the WFAP involves the authorization of access control when a MS is in the process of handed off to the WFAP from a macro BS. If the WFAP's CSG configuration disallows a certain MS, the MS cannot establish connection to the WFAP and will be dropped from coverage once the macro BS is no longer able to serve that MS.

The WFAP also needs to be pre-provisioned with several important parameters in order for the WFAP to support plug-and-play capabilities. These parameters are: device certificates, public keys for operator root CAs, FQDN of the bootstrap server, and a WFAP R6 identifier.

5.2.2 Security Gateway (Se-GW)

The Security Gateway provides the first entry into what is considered as the core part of the WiMAX ASN. WiMAX network architecture differs slightly from either the 3GPP or CDMA networks in that WiMAX's access part of the network also includes a gateway that would traditionally be considered as part of the operator's core network. However, because of the architecture difference as well as network model variation, the ASN-GW, and in the Femto case the Femto Gateway, is considered part of the access service network. The functional split of access service network and connectivity service network in the WiMAX architecture allows for faster deployment using many off-the-shelf network components and can accommodate different business case so that the operator can choose levels of investment in setting up the network that are easier for them and easier for other partners who also want to be in the game. Due to this difference in functional split, the first point of entry into the core port of the Femto ASN through the WFAP is the Femto Gateway via the Security Gateway. In addition, because of the use of SON server for self-organizing and auto-configuration features, the WFAP also connects to a SON through the Security Gateway. The purpose of the security gateway is to perform mutual authentication with the WFAP and to terminate IPsec tunneling for the Femtocell to other components of the network. The authentication is based on the certificate provisioned in the security gateway, which can happen during manufacturing phase or at a later time during deployment. Like the WFAP, the security gateway is also provisioned with one or more root CA public keys that are used to validate the device certificate in the WFAP as part of the mutual authentication. In addition, the security gateway also filters out unauthorized and unauthenticated traffic on the links between the security gateway and the WFAP. From the WFAP point of the view, anything behind the security gateway is considered the network, analogous to the core network in CDMA and 3GPP Femtocells. So the security gateway also performs access control function before admitting a WFAP to the network, for example, verifying that the WFAP's location is where the WFAP is allowed to operate (i.e. where the operator holds WiMAX spectrum license). Because of the R6-F split at the security gateway, when security gateway de-encapsulates packets that are destined for either the SON or the Femto Gateway, it verifies that the source identity of the packet is indeed the WFAP that initially established the IPsec tunnel with the security gateway. Though this is not a standard feature of the off-the-shelf security gateway, but it is important from a security perspective in the design of WiMAX Femto security that this is done so that any compromised WFAP cannot pretend to be another WFAP even after the establishment of the IPsec tunnel after WFAP and Security Gateway mutual authentication is done. As part of the modes of operation of IPsec, encryption and integrity protection are fully supported. Furthermore, as a standard feature of the security gateway, NAT traversal is also supported.

5.2.3 SON

Self Organizing Network functions include self-healing, self-configuration, and self-optimization (see Annex A.2). Though it is not something new, the WiMAX Femto fully utilizes the SON and incorporated it into its core Femtocell architecture. When the network is capable of detecting problems, mitigates and/or resolves the problems without any manual network administrator involvement, it is known as self-healing. On the other hand, self-configuration also refers to automatic configuration that can be thought of as plug-and-play. A WFAP is expected to be a plug-and-play device for the end user. After the user acquires the WFAP, the user is not expected to be able to take on the role of the network administrator to configure and deploy the WFAP before attempting to make the first call. Instead, WiMAX relies on a SON server that can help the user automatically configures and WFAP and therefore the only thing the user is required to do is simply provide the WFAP with power and Internet connection, powers the device on and the network would do the rest. Some of the SON features work in conjunction with each other to resolve issues. For example, if a path to the a particular network server is severed or becomes unreachable, the SON may be able to reconfigure the network path to avoid the bad path and reroute the traffic to the network using alternative paths, thus providing both self-configuration and self-healing at the same time. In another example, if the network detects that adjacent WFAPs are using frequencies that are close to each other or for some reason are interfering with each other, the network may increase the guard band between the frequencies that the two adjacent WFAPs are using or switch the frequencies of one of the WFAPs to an entirely different frequency to resolve the interference issues. This can be considered self-healing, self-configuration and self-optimization all at the same time.

The support of a SON server is not explicit mentioned in either the 3GPP or CDMA Femtocells because some of the SON functions may have been included with the Femto Management System within those networks. Nevertheless, the explicit SON support is a unique feature of the WiMAX Femto architecture.

Another function of the SON is to provide WFAP with location authorization because the WFAP is not allow to radiate on its air interface on the licensed spectrum without explicit authorization that it is within the legally allowed location where the operator holds the licensed spectrum. The process of location verification in WiMAX is quite similar to that of the location verification used in 3GPP Femtocells that can be found in Chapter 3. Another slight variation in the WiMAX Femto location verification is that SON can also rely on information from base station(s) or access point(s) from other wireless access technologies, such as Wi-Fi, CDMA, or GSM. This is because WiMAX access technology is envisioned to be a 4G service without any legacy access technology to fall back to. WiMAX also relies heavily on partnering with other access technology to support mobility and interworking so that in many cases, there is an underneath coverage of other type of access that WiMAX overlays on top it. In these cases, the SON is made aware of the presence of different access technologies present and can make use of it as part of the location verification for the WiMAX WFAPs. Another variation to point out is that manual intervention is also allowed where the WFAP user or subscriber is explicit asked to provide his location information, either via web interface or via a call from the SON operation center to confirm the user location. Though this is less reliable and subject to fraud, it is only used as the last resort when all else fails.

Since SON functionalities are considered management functions, some of these SON functionalities are split between the Femto Management System and SON server. For example, the FMS will be responsible for the upper layers of the self-configuration functions while SON server will be responsible for self-configuration of the air interface. Some of SON's self-configuration process involves functions such as location verification, automatic configuration system identifier and parameters, automatic neighbor discovery, auto-configuration of physical radio parameters.

Some of the SON's self-optimization functions include interference management, coverage and capacity optimization, mobility robustness optimization, mobility load balancing, dynamic reconfiguration management and measurement data collection. The measurement data collection function is used to gather measurement data that are used by the SON for self-optimization purposes. Usually, these measurement data are collected by the WFAP silently and sent to the network when network load condition is not heavy. These data include signal strength optimization of the WFAP neighbors, MS measurement reporting of WFAP radio scanning for neighbors, Femto HO parameters optimization and interference control, event measurements like cell specific call drops or handover failures, configuration data for pre-determine WFAP neighbor for a given MS, etc.

For details of how SON works and interacts with WiMAX Femto, the readers are encourages to seek out WiMAX Forum Network Architecture, detailed Protocols and Procedures Self-Organizing Networks. Additional information is also provided in the Annex section of this book (Annex A.2)

5.2.4 WFAP Management System

The use and support of a Femto Management System, called the WFAP Management System in WiMAX Femto architecture, is a standard feature in any Femtocells. The WFAP Management System is essentially an OAM (Operation, Administration and Management Server) that also supports other features. The OAM implements many network management protocols such as the SNMP, TR069 or DOCSIS that are commonly used partly to interface with other network elements and management servers in other networks to aid the operation and management of WFAPs in the WiMAX Femto systems. Similar to other FMSs, the WFAP Management System is also responsible for configuration of the WFAP as well as any software or firmware updates of the WFAPs.

5.2.5 Bootstrap server

Unique and explicit to WiMAX Femto architecture is the use of a bootstrap server. In the 3GPP Femto, some of the functions provided by the bootstrap server may have been depending on certain deployment. For example, in certain deployments of the 3GPP Femtocells, the operator also deploys a provisioning security gateway for the FAPs to initially access. The location and address of the provisioning security gateway may have been initially installed in the factory or at operator deployment site. Upon connecting to the provisioning security gateway, the FAP then gets the IP address of the serving security gateway and can proceed to connect to the operator's core network. The bootstrap server in the Femto ASN also supports initial bootstrap procedures for the WFAP, which also include a secure tunnel setup and an authentication by the WFAP of the bootstrap server by validating the certificate presented to the WFAP in the secure connection establishment process. The certificate pre-provisioned in the bootstrap server is used for this authentication. The WFAP bootstrap procedures involves connecting to the bootstrap server for downloading initial configuration information based on the WFAP's IP address, location information or other information necessary for the bootstrap server to select the appropriate serving security gateway.

5.2.6 Femto-AAA Server

This network entity maintains the WFAP subscription information, performs authentication, authorization and accounting for the WFAP. Since WiMAX operator can have the flexibility and a variety of different billing and charging models depending on who is providing the access service and who is providing the connectivity service, the Femto-AAA is fully capable of supporting various billing models.

5.2.7 Femto Architecture Reference Points

The various network elements are connected in a series of reference points as below:

- R1: provides the interface between the SS and the BS
- R2: provides the interface between the SS and the CSN
- R3: provides the interface between the ASN and the CSN
- R4: provides the interface between an ASN and another ASN
- R5: provides the interface between a CSN and another CSN
- R6: provides the interface between BS and the ASN-GW

- R7: provides the interface internally within the ASN-GW of the ASN
- R8: provides the interface between two BSs within an ASN

To support Femtocells, the following reference points are further defined or extended to include Femto components:

- R3-F: provides the interface between the Femto ASN and the Femto CSN
- R4: provides the interface between a Femto ASN and a macro ASN
- R6-F: provides the interface between the FAP and the FGW
- RbF: provides the interface between the FAP and a bootstrap server located in the Femto CSN

The reference points mentioned above for the macro network as well as in the Femtocells cover both control plane and bearer plane. The bearer plane is sometimes called data plane or user plane in CDMA or 3GPP and they are essentially the same for carrying user data over the reference points.

Reference point R1 is unchanged from a MS's perspective as it sees only a base station, whether it is a macro base station or a WFAP, though the MS may have enhanced capability if it is capable of supporting some of the more advanced features associated with the Femtocells such as subscriber group access control. The R3-F reference point connects the Femto ASN and the CSN to support AAA and to carry user data. The R6-F differs slightly from its counterpart R6 in the macro ASN in that the R6-F control plane gets split into two components, one component supporting secure tunnel over IPsec between WFAP and the SeGW and the other component supporting normal IP tunnel between SeGW and the FGW. The part of the R6-F over IPsec between the WFAP and the SeGW is a normal R6 that is encapsulated over IPsec (i.e. WiMAX header over UDP over IP for the control plane traffic) and the part of R6-F between SeGW and the FGW is essentially identical to a normal R6. This applies for both control plane traffic as well as over bearer plane traffic, except in the case of bearer plane traffic, GRE or Generic Routing Encapsulation is used over IP over IPsec instead of WiMAX header over UDP over IP over IPsec. In retrospective, the different layering of encapsulation in WiMAX, one may think that it may not be as efficient as some of the other access technologies such as GSM or CDMA, but because of the anticipated data rate of the WiMAX services, tens of bits of extra overhead in a multi-million bits per second scheme really becomes a moot point where in the original CDMA air interface design, every bit of efficiency on the air interface could mean huge potential efficiency gains.

5.3 WiMAX Femto Security Features and Mechanisms

The security mechanisms and procedures in the WiMAX Femtocell include the following:

- WFAP Initialization and authentication
- WFAP Network Exit
- CSG Management
- MS operations
- Management Procedures
- Accounting
- WFAP backhaul fault detection and mitigation

These mechanisms and procedures, though similar mechanisms and procedures have been described in greater details in Chapter 3 for the 3GPP Femtocells and Chapter 4 for the CDMA Femtocells, are used for the secure operations of the Femto systems in WiMAX. There are, of course, differences, which are described in more details later in this chapter. These differences are mostly contributed more to the different network architecture and model rather than to the difference in security philosophy.

5.3.1 WiMAX Femto Access Point Initialization

As previous mentioned, WFAP are pre-provisioned with several important parameters before the WFAP are deployed in a WiMAX network, namely, device certificate, public keys of the operator root CAs, FQDN or IP address of the bootstrap server, and an R6 Identifier.

Like the other Femtocells based on 3GPP or CDMA access technologies, WiMAX Femtocells also use a certificate-based authentication. In the certificate-based authentication scheme, the WFAP presents its device certificate to the security gateway and the security gateway presents its certificate to the WFAP. The two entities validate each other's certificate based on the entity that issued the certificate for these entities. In the WFAP, a pre-provisioned X.509 device certificate and the public keys for the root CA that are used for validating an equivalent X.509 certificate in the security gateway and/or in the bootstrap server. This pre-provisioning process may occur during the WFAP manufacturing stage or some other stage of the initial deployment, such as during initial network service provider setup. The entity (i.e. root CA) that issues the Se-GW's X.509 certificate and bootstrap server's X.509 certificate may be the same or may be different, depending on operator requirements or other factors. When they are the same, the same public key of the root CA that is pre-provisioned in the WFAP is used to validate Se-GW and bootstrap server. If there are two root CAs involved in the issuance of two different X.509 certificates in the Se-GW and bootstrap server respectively, which is a distinct possibility given that there may be two network providers that operate the WiMAX's ASN and CSN independently due to the flexible architecture in WiMAX, then there need to be separate public keys from each root CA respectively pre-provisioned into the WFAP.

During initialization, the WFAP boots up, goes through its internal procedures. Though not explicitly called for or stated in the specifications, it is also likely that the WFAP goes through a device validation process similar to that of the 3GPP or CDMA Femtocells. Since the validation process, if it exists and given that the WFAP Management Server's support for TR-069-based protocols, it would be natural to assume that the process indeed exists, would be quite similar, details are left out here. The WFAP needs to establish communication with the bootstrap server to get initial system parameters. There are several ways in which the WFAP discovers the IP address of the bootstrap server, for example using DHCP and/or DNS servers or using manual static configuration during setup, since this parameter can also be pre-provisioned prior to the WFAP's initialization procedure. The WFAP authenticates the bootstrap server using the bootstrap server certificate presented to the WFAP using HTTPS and in the process establishes a secure connection between WFAP and the Bootstrap server over SSL or TLS. IPsec can also be used alternatively. The authentication between WFAP and bootstrap server is only one-way in the sense that the WFAP authenticates the bootstrap server and the bootstrap server does not authenticate the WFAP. This secure connection to the Bootstrap Server is needed whenever the WFAP needs to locate the Se-GW, Femto Gateway, (possibly) the SON (if it is not to be provided by the WFAP Management System), and the WFAP Management System, such as when a new WFAP is put into service or when the WFAP Management Server address or Se-GW address has changed. It is entirely possible that in some static deployment environment, the WFAP Management Server information may also be pre-provisioned into the WFAP if it is not expected to change throughout the lifetime of the WFAP's deployment. However, it would be much easier to use the bootstrap server for this process since this initialization process is explicitly defined in the WiMAX Femto Systems. Additionally, it is also entirely possible that the Femto Gateway is being selected by the security gateway depending on configuration and operator policy.

Once WFAP is armed with the addresses of the Se-GW and WFAP Management Server, the connection to the bootstrap server is no longer required and the secure tunnel is torn down. Subsequent power-on and power-off cycles of the WFAP would be made without going through the bootstrap server again as the sole purpose of the bootstrap server is to serve the provisioning of the WFAP with parameters that are necessary to set up service locally within the local serving network.

The next step in the process is the mutual authentication between the WFAP and the Security Gateway using IKEv2 authentication and key establishment procedures and the x.509 certificate pre-provisioned in the WFAP. Details of the IKEv2 authentication and key establishment procedure are essentially identical to that of the procedures used in CDMA and 3GPP Femto authentication schemes. As a result of the successful authentication and key exchange, IPsec tunnel and IPsec Security Associations are established between the WFAP and the Security Gateway. An example of a 3GPP Femtocell IKEv2 procedure and call flow can be found in Chapter 7. Since the IKEv2 call flow is essentially identical for all of the certificate-based authentication procedures in various Femto architectures, a WiMAX-specific IKEv2 call flow will not be included separately.

To further the initialization process, the Se-GW also performs a WFAP authorization process by contacting the Femto AAA in order to verify that the WFAP is authorized to provide Femto service (e.g. Femto subscription information about a particular WFAP). Next, the WFAP connects to the WFAP Management System to get latest WFAP configuration parameters from the WFAP management system. WFAP sends its local information to the WFMS which also includes its IP address, HW S/N, S/W version, and location information and in returns receives higher layer configuration parameters and either the FQDN or the IP address of the SON server. Since the SON server FQDN or IP address can also be provided by the bootstrap server, the WFAP–MS may not need to provide that information here at this configuration download process.

Next the WFAP connects to the SON to can obtain additional information and parameters and allow the SON to perform additional authorization so that the WFAP can begin service. Some of the additional information includes location information where the WFAP is authorized to use its radios (i.e. where the network provider has spectrum license to operate WiMAX service) and radio configuration data. As part of the configuration data, it may include frequency to use, neighbor information, etc. If the Femto GW provided by the Security Gateway is not the optimal serving Femto gateway for the WFAP based on its current location, for example, the initial authentication in the security gateway may have netted a serving Femto Gateway that normally does not serve where the WFAP is located or even the authentication and IPsec tunnel establishment made was not with the optimal serving security gateway, the SON server may also provide the optimal or preferred Femto Gateway's identity, along with R6-ID of the WFAP, and/or the ID of the Se-GW that is associated with the Femto Gateway to the WFAP as part of the exchange of parameter with the SON server. If the security gateway used to connect to SON or Femto Management System was not the optimal security gateway, the IPsec tunnel between the WFAP and the security gateway will be torn down and a new IPsec tunnel established with the correct serving security gateway. At that point, the WFAP also registers with the Femto Gateway.

The WFAP may also use the available DHCP option or DNS server in the network to locate a particular server, whether the server is a SON server, the security gateway, a Femto Management System based on a FQDN pre-provisioned, or information regarding the server returned as part of the initialization process.

In the process of WFAP security gateway authentication and IPsec tunnel management, the significance of the WFAP R6 identifier is for the Femto GW to know where the R6 is terminated since R6-F is split between the WFAP-to-Se-GW and Se-GW-to-FGW. WFAP packets destined to the FGW are encapsulated in the IPsec tunnel and before the security gateway strips off the encapsulated IPsec header, it needs to know about the R6 identifier of the WFAP in order to perform integrity validation on the R6 control packet since it is required that the security gateway needs to verify the source of the R6 packet is indeed coming from the WFAP which established the IPsec tunnel. Additionally, the security gateway may also perform further checks on all R6 control messages to ensure the correct R6 identifier is used before messages are passed onto the Femto gateway. The R6 identify used by the WFAP can be either a MAC address or an IP address. When MAC address is used for this R6 identifier, it could be the MAC on the LAN or WAN interface, the BS ID, the X.509 device certificate identifier, etc. and it does not have to be the same value used in the WFAP's X.509 device certificate. When this R6 identifier is not pre-provisioned in the WFAP, the security gateway may choose the IPsec tunnel inner IP address as the identifier for verification purposes.

- WFAP Initialization and authentication
- WFAP Network Exit
- CSG Management
- Mobility Management
- Accounting
- WFAP backhaul fault detection and mitigation

5.3.2 WFAP Network Exit

After WFAP's initialization, setting up of the IPsec tunnel and registration with the Femto Gateway, the WFAP is ready for service in terms of accepting MS into the network. At times, the WFAP may need to be taken out of service or disconnect from the network, for example for maintenance purposes. WiMAX defines an explicit WFAP network exit process which the WFAP may be taken off the network access gracefully such as the case of user or network initiated procedure or it may be taken off the network ungracefully or abruptly, for example when power suddenly fails.

When a graceful network exit occurs, the WFAP or the network triggers the network exit procedure while the backhaul connectivity is still intact between the WFAP and the Femto ASN. In this case, the WFAP will stop admitting new MS into the network (e.g. MS that are being handed off to the WFAP) but will continue to serve existing MSs that are already being served by the WFAP (i.e. that have active sessions). As part of the graceful network exit, the WFAP goes through normal connection tear-down process for each of the servers that it has a connection: the SON server, the WFAP management server, the Femto Gateway, and the Security Gateway respectively in that order. Graceful network exit also helps to speed up the process when the WFAP attaches to the network subsequently when the WFAP is able to save certain useful parameters or states in the process of tearing down the connections with various servers.

Ungraceful network exit occurs when Fe-GW triggers the WFAP network exit process, such as in the case that Fe-GW is unable to maintain network connectivity with the WFAP for example when the WFAP does not answer the Fe-GW pings for a period of time. The network connectivity with the WFAP may be lost due to a number of reasons such as power is lost to the WFAP or the WFAP loses IP connectivity. When the Femto Gateway has no knowledge of what state the WFAP is in, the only think it can do is to trigger the network procedure by asking the various servers to mark the network connections (i.e. secure tunnel) with the WFAP invalid. The process may also involve clearing cache in these servers. Even when the WFA restores power and attempt to reconnect based on the state information it maintains with the various servers, the servers would have to request that new connections to be established instead of attempting to restore the old connections as those connections on the server side would have been marked invalid or expired.

5.3.3 Closed Subscriber Group Management

CSG management is both a security function as well as an authorization function on both the network side and on the MS side. Since WiMAX Femtocell is designed to support both legacy MS and Femto-aware Mobile, CSG verification in either the MS or in the WFAP is supported. In the network-based CSG scheme, a white list (i.e. MSs that are explicitly allowed to be admitted) or a black list (i.e. MSs that are explicitly excluded) may be used. In the MS-based scheme, which requires the use of Femto-aware MS, either a white list or black list of WFAPs that the MS is or is not a member of is stored. Femto-aware MS may get and/or update the CSG list in one of several manners, OTA, manual selection, etc. The CSG admission process is based on the information in the CSG whether the MS is allowed to access the WFAP, and is implemented as part of the MS network entry process for MS accessing WFAPs that have a closed access mode (called CSG-Closed WFAP in WiMAX). CSG also plays an important part in other MS or network related procedures such as location update, paging and idle-mode handling.

5.3.4 Mobility Management

Mobility of MS is a big part of WiMAX strategy, which may involve MS moving between a WFAP to a macro base station, between macro base stations, and between WFAPs. Mobility events can be triggered by either the MS or by the WFAP when certain handover condition is reached. Part of the decision whether a particular WFAP is allowed to receive a MS depends on the CSG status. In the case of MS supporting the CSG capability, the MS may scan only the member of the WFAPs that is allowed to receive it and skip any other WFAPs that are not configured to allow the MS access. This CSG consideration is applicable when the target base station is a WFAP and does not apply when the target is a macro base station. Macro base stations can also implement CSG functionalities but is very unlikely as it will no doubt increase the operational complexity as well as management complexity of the system.

5.3.5 Network Management Procedure

Other network related procedures, though not considered as part of security procedures and mechanisms explicitly, are necessary part of the overall secure operations of the WiMAX Femtocells. These include QoS control, accounting, WFAP backhaul fault detection and mitigation procedures. These procedures and processes are performed as part of the operation of the WiMAX Femto services and involve various servers in the network, including the SON server, the Femto Management server as well as the Femto-AAA server.

5.4 Differences between WiMAX and 3GPP/CDMA Femtocells

Many of the differences between the WiMAX Femtocells and the 3GPP Femtocells have been described in one way or another when the specific security features, security mechanisms and security procedures are discussed in details in the previous sections. For the most part, the differences are also applicable to the CDMA Femtos. In fact, the CDMA Femtos have more commonality to the 3GPP Femto than the WiMAX Femto. This section serves as a summary to capture the differences. As the differences between the 3GPP and CDMA Femtos have been discussed in more details, some of the details will be left out here.

5.4.1 SIM Card Support

The current versions of the specifications for the WiMAX Femtocells do not support the use of SIM card, either mandatory or option even though certain versions of the WiMAX MS do support the use of SIMs. There is no information on whether future versions of the WiMAX Femtocells will support the SIM card.

5.4.2 Use of Pre-shared Key

The WiMAX Femtocells are designed from the very beginning to be used without a pre-shared key. In this case, there have been no known case of WiMAX Femto deployments where the WFAPs are developed prior to standards, which means there are no known existence of pre-standards WFAPs that may have used pre-shared keys for authentication as in the case of either the 3GPP or CDMA Femtocells. This is a cleaner approach to system deployment and interoperability issues have been avoided.

5.4.3 Use of Device Certificate

Current specifications in all three access technology dictate the use of device certificates for use in the authentication and setting up of the IPsec tunnel between the WFAP and the Security Gateway. Only the 3GPP Femtocells support the automatic certificate enrollment procedure to replace the manufacturer's device certificate with an operator certificate while WiMAX Femtocells do not.

5.4.4 Optional Use of IPsec

IKEv2 is used as the authentication and key exchange protocol for setting up the IPsec tunnel and Security Association between the FAP and Security Gateway. However, in the 3GPP networks, the use of IPsec tunnel between the FAP and the Security Gateway is optional while it is mandatory in WiMAX Femtocells.

5.4.5 Direct Interface between FAPs

WiMAX Femtocells do not support direct interface between the WFAPs for mobility purposes. In the 3GPP Femto, the direct interface is being used in some Femtocells to aid faster handoffs between FAPs, especially in enterprise scenarios and in large residence deployment scenarios where MS/UE handoffs are considered frequent. The support of operator enrolled certificates in 3GPP makes the support of direct interface possible where the FAPs can easily authenticate each other and setting up the IPsec tunnel based on commonly issued certificates. Though this direct interface feature may be easily supported in WiMAX Femto as well, as of now there is no information about its current or future support. No future known enhancements are expected at this time.

5.4.6 CSG Handling

CSG lists in WiMAX Femtocells can be either a list of MSs that is allowed to access a WFAP or a list of WFAPs that is a MS is allowed to access. Both the 3GPP Femtocells and the CDMA Femtocells seem to support only the former.

5.4.7 FAP Identity Verification

There is FAP identity verification in both 3GPP Femtocells and in WiMAX Femtocells. In WiMAX, this identity is explicitly a R6 interface identity. This R6 identity can be a pre-provisioned identity, an IP address, or a MAC address of the WFAP. In the 3GPP Femto, the

FAP identity can be the FAP device identity, a CSG identity or other equivalent identity (e.g. connection ID) as long as that identity can be uniquely mapped to the FAP. Another difference is that in the WiMAX case, the verification explicitly takes place in the security gateway where it needs to strip off the outer header of the IPsec encapsulated packet before handing it off to the next node (i.e. Femto Gateway) where in the 3GPP case, the where and how verification is done is left for the discretion of vendor implementation, such as:

- An interface between SeGW and FGW, using an already standard protocol;
- Authentication function in FGW;
- Authentication function in other network entities (e.g., HLR in case of USIM based authentication);
- Using an already standard protocol between FGW and an entity (e.g., DHCP Server) that holds the allocated IPSec address.

5.4.8 LIPA Access

Since WiMAX is intended to be a high speed broadband access, LIPA access is not expected to be necessary and thus is not explicitly covered by the standards in WiMAX.

5.4.9 Security Profiles

Other differences and variations in which the certificates, IPsec, TLS, and IKEv2 are used between WiMAX, CDMA, and 3GPP Femtocells do not affect the operation of these Femtocells, for example, the use of EAP-TLS is supported in WiMAX explicitly whereas it is not explicitly forbidden in other Femtocells.

5.4.10 Other Differences

There are also other differences among the three Femtocells based the WiMAX, CDMA and 3GPP that are attributed to the differences in fundamental designs and operations of these three distinct access technologies. Interested readers are encouraged to further explore those differences in greater details via other avenues.

6

LIPA and SIPTO

Recent study shows that in the EU, over 70% of the mobile data traffic is routed over Wi-Fi. Even though offloading is not part of the overall WiMAX Femto strategy, LIPA and SIPTO is greater part of making the CDMA and 3GPP Femtocells attractive to the operators and users. LIPA, or Local IP Access and SIPTO, or Selected IP Traffic Offload both offer tremendous opportunities for the user to experience greater access to the Internet in terms of speed and through put and at the same time for the operator to reduce data traffic bottleneck in an effort to enhance user experience.

LIPA provides access for IP capable UEs connected via the Femto Access Point to other IP callable entities within the same residential or enterprise IP network. Some use cases of LIPA include accessing an IP-capable printer, an IP-capable camera or IP video server from the user's mobile equipment while user is connected to a FAP. In most cases, with the ubiquitous use of Wi-Fi access points in residential and enterprise environments and Wi-Fi capable user equipment available, some of the LIPA's original use cases may become a moot point in the future, especially when all of the IP-capable entities are expected to be accessible using the available Wi-Fi access point. This creates both a challenge and a dilemma for the operator as the operator may experience potential revenue loss due to the fact that the operator's core network is no longer being used for the local access.

SIPTO differs from LIPA in the sense that when a MS connected to a FAP that is accessing a particular IP network, instead of routing the IP traffic via the FAP and through the operator's core network and then through one of the operator's IP gateways (e.g. Serving Gateway) to the public Internet, SIPTO allows a more direct access for the MS to the Internet thereby bypassing the operator's core network. As mentioned previously, this may be beneficial to both the user and the operator. As

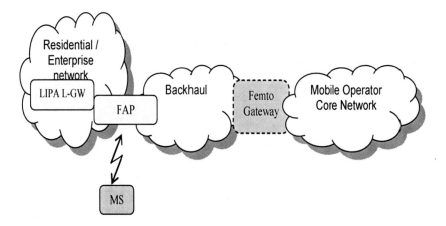

Figure 6.1. Femto LIPA Architecture

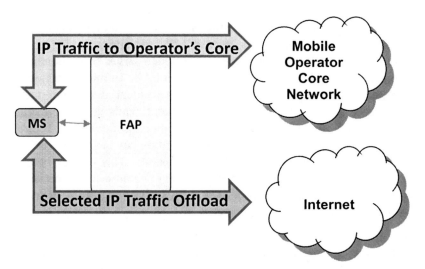

Figure 6.2. Femtocell SIPTO Architecture

the user has a more direct connection to the Internet, the user's Internet speed would be faster and therefore user's satisfaction is increased as a result of the faster connection (i.e. faster content delivery). While the scope of the offload in Femto-based architecture is different than the access point or core network based offload, the intention is the same—to increase IP traffic throughput and to minimize network load at the core network.

Besides the clear benefits, LIPA and SIPTO also present some unique security challenges to the operator.

6.1 Security Considerations in LIPA and SIPTO

The benefits of the LIPA and SIPTO are clear, but the considerations for their use are strictly based on operator policies as there would be both security and regulatory implications. These considerations are mostly security related as the users' traffic is no longer visible in the operator's network once it is offloaded via LIPA or via SIPTO and there are multitude of these considerations from Legal Intercept perspective, Billing and Charging perspective, and Air Interface security perspective respectively.

6.1.1 Legal Intercept

Legal Intercept, also commonly known as lawful intercept, is both a security and regulatory requirement to consider in LIPA and SIPTO. As the user traffic is routed to either the local residential or enterprise network in the case of LIPA or to the public Internet in the case of SIPTO, the operator still needs to comply with regulatory requirements to provide legal intercept should such request be made by a law enforcement entity. Traditionally, legal intercept function is provided on the network side by an Administration function and a Delivery function. The Administration function interfaces with the Law Enforcement Monitoring Function or LEMF that sends request to intercept to the 3GPP network. The intercept request includes information about the specific MS to intercept, the valid duration of the request, etc. The Administration function sends the request to the operator's network. The appropriate network element that hosts the legal intercept functions captures the Intercept Related Information (IRI) and Communication Content (CC) and sent them back to the law enforcement agency via an interface between the Delivery function and the LEMF. If the communication as requested by LEA is ciphered, the mobile operator can either deliver the ciphered content along with the ciphering key so that the LEA can decipher the content or the operator may decipher the content before delivering to the LEA. The communication content may include voice, data, video, SMS, or media formats that is supported in the operator's network.

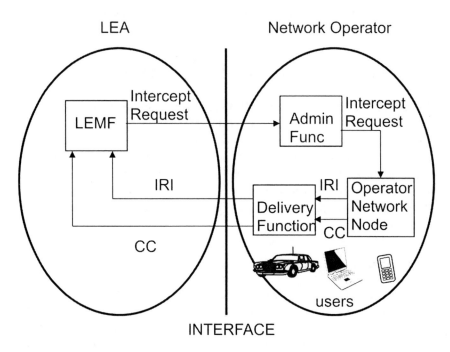

Figure 6.3. Legal Intercept in Conceptual Architecture

In case of normal traffic routed via the operator's core network, the legal intercept can be easily achieved since the content of the user communication traverses entirely within the network elements that are directly controlled by the operator, whether they are the base station or the network gateway nodes. And because the legal intercept request has to be served discretely without the user being watched becoming aware about it, from the user's point of view, the network does the intercept silently. In the case the base station is a Femtocell, there is no difference between how it is handled in the network as the Femto Access Point is also considered as network equipment. If the Femto Access Point implements the LI function, it may capture IRI and CC as per the LI request and silently send the information to the network. However, when the traffic is LIPA-based or SIPTO-based, the operator's core network is completely bypassed, especially for the traffic that the user originates although mobile terminated traffic may still traverse the operator's core network in some cases, for example if the user is access operator owned servers or contents. If the operator is not able to access the content, it cannot fulfill the LEA request to provide IRI and CC of the intercepted subscriber. Since the operator has to comply with the legal intercept requirements, the operator has to provide a solution to satisfy these requirements.

Operator policies on SIPTO and LIPA may be used by the operator. Since it is an operator option as to when to enable the LIPA and SIPTO use for a particular subscriber, the operator may choose to disable the use for a particular subscriber if there is an active legal intercept request. The user in this case should not suspect that he or she is being targeted for legal intercept as LIPA and SIPTO are designed to be transparent from the user's perspective without the need of user intervention. Another word, the user does not choose to turn on or turn off LIPA or SIPTO. It is controlled by the network based on the network operator's policy. In addition to supporting legal intercept in the policies, the policies may also support other information pertaining to a particular's user traffic, for example when to offload a user's traffic and what type of traffic to offload.

6.1.2 Billing and Charging

When LIPA or SIPTO is enabled, user traffic that does not traverse the operator's core network may not be properly accounted for by the operator. Billing and charging the users become a potential issue for the operator. On the other hand, if the operator's core network is not being used to route user traffic, would the user still be responsible for paying the operator for the amount of data sent and received (assuming that the user is not on a fixed all-you-can-eat data plan)? Answer to the above question may present a dilemma to the operator regardless of yes or no. If user needs to pay for the amount of data even though the operator's core network is not used to carry some of the traffic, how much should the user be responsible and how would the operator be able to generate usage? If the user does not need to pay for the amount the data not routed via the operator's network, the operator may risk the potential of revenue loss due to the fact that less traffic now flows through the operator's network. How the operator handles this billing and charging issue then becomes an important topic for the operator to decide whether to enable LIPA and SIPTO or not. One solution is for the operator to charge per usage not based on the amount of data. Another solution is to charge a flat fee for using LIPA and SIPTO as a value added service. If the operator decides that it should charge the subscriber based on actual data usage, then the operator need to capture the amount of data correctly to bill the user. This then may potentially turn into a security issue. Since the operator has no way of knowing the amount of data being offloaded, the operator may not be able to capture it correctly so it may have to rely on the MS to report it. Another word, the operator may install a counter or a billing collection agent in the subscriber's equipment to report to the network after each LIPA or SIPTO session completes. If the user's equipment is tampered with, there is little or no security guarantee that the usage data as being reported by the user is accurate, especially since the user would have financial incentive to under report the usage in an attempt to get a lower bill.

6.1.3 Backhaul Security

When the user traffic is routed via the operator core network, the FAP is used as the first point of entry into the operator's core network. Furthermore, the FAP and the core part of the operator's core network are protected via the IPsec tunnel established between the FAP and the Security Gateway as a result of the mutual authentication between these two network elements. Security is provided by encryption on the air interface from the MS to the FAP, while security to the core network is protected by backhaul security (e.g. NDS/IP that is specified in 3GPP based on the use of IPsec). However, when the user traffic no longer goes through the backhaul, there is no protection on the backhaul and the protection may need to be relied upon the security of upper or intermediate layers (e.g. application layer, transport layer or even IP layer), even though the L-GW and the core network still needs to maintain control signaling that is still routed via the IPsec tunnel. From the users' perspective, there is no guarantee that the upper or intermediate layers provide the same level of security as provided over the backhaul for their traffic. As more and more emphasis has been placed on end-to-end security rather than relying on hop-by-hop security, backhaul security may add to the peace of mind that the users have traditionally expected to have and any security on top of the backhaul security would be an added bonus for the users.

6.1.4 Mobility Support

When direct interfaces between the Femtocells are not supported or are not implemented (e.g. WiMAX- or CDMA-based Femtocells), mobility support of MSs would mean requiring sub-optimal routing of MS's traffic back to the operator core. Security contexts between the Femtocells would have to be exchanged via the operator core in order to support any mobility events when the MSs are offloaded via LIPA or SIPTO. Increase of network activities related to the exchange and transfer of security contexts may unnecessarily complicate system operation and if not handled properly may lead to exposure of user security or privacy.

6.1.5 L-GW to Femto Interface

The addition of the L-GW to the network also requires the interface between L-GW and the Femto to be secured. Current Femto architectures in 3GPP support that both a co-located L-GW and a stand-alone L-GW model. When co-located L-GW is supported, the interface between the L-GW and the Femtocell is hidden from the outside. Since the L-GW is considered as a separate logical entity, even if it is co-located with the Femto Access Point, the interface between the two logical entities still needs to be secured. Since most Femtocells require that a hardware-based root of trust be part of the system that guarantees the physical and platform security of the FAP, the L-GW can be viewed as another module that need to be incorporated in the trusted environment. Another word, when the FAP powers up, the L-GW module needs to be validated before being loaded and put into service. The L-GW may also require a separate IP address to be allocated by the security gateway apart from the IP address that is allocated to the FAP at time of IPsec establishment. When the L-GW is not co-located with the FAP, the interface between the L-GW and FAP should be IP-based and can be provided by IPsec or TLS. However, this would also require that the L-GW be provisioned with a certificate (e.g. vendor device certificate or operator enrolled certificate.

6.2 In Short

LIPA and SIPTO are well supported in all Femtocell architectures and are able to provide great benefits to the operator and users alike. Although the security issues discussed here are not new, but they are real and they need to be addressed one-by-one. Fortunately, these security issues are not insurmountable that would prevent the operators from taking advantage of the offload opportunities, both for the users and for the operators themselves. The vendors and the operators have come a long way and have great deal of experience providing solutions to resolve issues just like the ones discussed here. When the issues and concerns are properly addressed, indeed LIPA and SIPTO can be a great tool for the operators and users alike.

7

Security Profiles and IKEv2 Call Flow

7.1 Security Profiles

Many system designers and implementers choose to reuse the set of rich security protocols that have been researched, standardized, implemented, and proven secure against a variety of attacks and in variety of environments. Many of these protocols, algorithms, and tools have gone through the standardization process such as in IETF or other standards with the necessary security analysis. Reusing these well defined protocols makes perfect sense. However, it is crucial to ensure that the security protocols, algorithms, etc. are used securely and properly in the way that they are originally intended to operate when these protocols and adopted into environment that they may not have been designed to work. Security Profiles plays an important part of making sure that the security protocols that are used securely and properly. Although many of these security protocols are flexible enough that they can accommodate a variety of options, ciphering algorithms, integrity algorithms, and operating modes, so on so forth. In case of 3GPP, not only are the security protocols profiled, the certificates used (device certificate for device authentication and TLS certificate for FMS authentication) are also profiled. It is important that the profiles and usage guidelines are followed properly. Along with the detailed security profiles (based on 3GPP interpretation of the Femtocell), including that of the profile of TLS and IKEv2, an example of IKEv2 call flow is also given, further illustrating how profiles are used.

7.1.1 TLS Certificate Profile

In 3GPP Femtocells, the Femtocell and the FMS use certificates for mutual authentication and establishment of TLS security tunnel. IETF RFC 5280 already has an extensive list of options and uses for certificates. But the use of these certificates as Femtocell and FMS certificates has the following restrictions on public key size, algorithms to support and to use for certificate signature, CRL extension and handling, as well as subject and issuer name format that differ from or in addition to IETF RFC 5280. IETF RFC 5280 has been around since May of 2008 and was designed to support many signature algorithms and public key sizes. Since then, some of these algorithms have become outdated and replaced by stronger algorithms or with key sized. In particular, the SHA-1 algorithm was published in 1995 by NIST as the hash algorithm of choice for many message authentication and signature calculation applications and has proven both useful and reliable. However, recent advances in both computing capability and research have yielded potential weakness in SHA-1. NIST has since recommended that SHA-1 be longer used and replaced with a stronger algorithm with larger block size SHA-256. But since there are still many older versions of the certificates in use, and for obvious compatibility reasons, these older versions of certificate are still supported in one way or another. A clean slate would go a long way for improved security when it is done properly. For that reason, some of these restrictions are put on the certificates being used for the Femtocell:

- Only version 3 certificate as defined in IETF RFC 5280 is used

- For signing certificate, SHA-1 is only used for early release Femtocells (i.e. pre-Release 9) and SHA-256 is used for all other releases in signing certificates. MD-5 cannot be used. And MD2 can only used for backward compatibility reasons.

- Only RSAEncryption is used as signature algorithm and rsaEncryption as public key algorithm

- Public key length is 1024 for backward compatibility reasons and 2048 for all newly created certificates

- CRL is restricted to version 2

- CRL retrieval with LDAPv3 as mandatorily supported method while HTTP may also be supported and used.

- The Femtocell TLS certificate shall be signed and issued by operator authorized CA, which can be operator CA, vendor CA, or a third party CA

- The Femtocell TLS certificate shall use FQDN format for its unique identity in both the subjectAltName extension of type dNSName and in the common name field

- Similarly, the FMS TLS certificate shall use FQDN format for is identity in both the subjectAltName extension of type dNSName and in the common name field

- Femtocell TLS certificate may be enrolled to an operator PKI using procedure defined in 3GPP

- The public key length shall be at least 2048-bit and a public key length of at least 4096-bit shall be supported for the CA certificate used in TLS

7.1.2 IKEv2 Certificate Profile

The Femtocell certificate and Security Gateway certificate used in device authentication and IPsec tunnel establishment have similar requirements as the TLS certificates. In addition to a number of requirements already specified in 3GPP TS 33.310, the following additions and exceptions also apply:

- The Femtocell IKEv2 certificate shall be signed and issued by operator authorized CA, which can be operator CA, vendor CA, or a third party CA

- The Femtocell IKEv2 certificate uses FQDN format for its unique identity in both the subjectAltName and this identity shall be the same as the identity used in IDi payload of the first IKE_AUTH request during the IKEv2 process (see IKEv2 call flow)

- Similarly, the FMS TLS certificate uses FQDN format for is identity in both the subjectAltName extension of type dNSName and in the common name field

- CRL and/or OCSP may be supported and information regarding these servers may be configured in the security gateway or carried in the security gateway certificate.

- The public key length shall be at least 2048-bit and a public key length of at least 4096-bit shall be supported for the IKEv2 CA certificate. In addition, there is no restriction in the issuer name for both the Femtocell CA certificates as well as the security gateway CA certificates.

7.1.3 TR-069 Protocol Profile

FMS is an important part of the Femtocell architecture and supports a plethora of Femto management functions, such as Femto configuration storage (example of some security related parameters are given in Annex A) and software/firmware upgrade capabilities. The FMS can almost be considered as an off-the-shelf component within the Femto system and is implemented using the CPE WAN Management Protocol TR-069 as defined by the Broadband Forum. However, some use cases may not be entirely applicable for the Femtocell and some restrictions and extensions are put in place on the use of TLS, authentication methods, ciphersuite support, etc.:

- TLS profile shall follow the profile defined in normative part of 3GPP TS 33.310

- Shared-secret-based authentication between Femtocell acting as CPE and FMS acting as ACS shall not be allowed. Only certificate-based authentication shall be allowed.

- The use of TLS to transport the CPE WAN Management Protocol shall be mandatory in case that the FMS is accessible on public internet or when TLS is used within the IPsec tunnel.

- The FMS URI shall be specified as an HTTPS URL in case that the FMS is accessible on public internet or when TLS is used within the IPsec tunnel.

- Ciphersuites with RC4 shall not be used. The support of TLS cipher suite RSA_WITH_RC4_128_SHA shall not be mandatory

- The Femtocell acting as CPE shall not be obliged to wait until it has accurate absolute time before it contacts the FMS acting as ACS.

- If the Femtocell contacts the FMS without having the accurate absolute time (i.e. UTC), it shall not ignore components of the FMS certificate related to absolute time, e.g. not-valid-before and not-valid-after certificate restrictions, but use a local clock instead even when the local clock may not have advanced due to the FAP having lost power between power cycles.

- The support for FAP authentication using client-side (CPE side) certificates shall be mandatory.

- The FAP acting as CPE shall be authenticated to the FMS by the FAP identity contained in the FAP certificate in case that mutual authentication between FAP and FMS is performed.

7.1.4 IKEv2 Usage Profile

Since the use of certificate is extensively defined in the Femtocell, the use of pre-shared secret for authentication is neither supported nor allowed, even though IKEv2 supports such modes of operation. The H(e)NB and the SeGW shall conform to the profile of IKEv2 as specified in clause 5.4.2 of TS 33.210 [9] with the exception that the use of pre-shared secrets for authentication is not supported.

The following additional requirements on certificate based IKEv2 authentication for the IKE_INIT_SA and IKE_AUTH exchanges shall be applied:

- The use of RSA signatures for authentication shall be supported.

- The H(e)NB shall include its identity in the IDi payload of the first IKE_AUTH request.

- The H(e)NB identity in the IDi payload may be used for policy checks.

- Initiating/responding end entities are required to send certificate requests in the IKE_INIT_SA exchange for the responder and in the IKE_AUTH exchange for the initiator.

- The messages for the IKE_AUTH exchanges shall include a certificate or certificate chain providing evidence that the key used to compute a digital signature belongs to the identity in the ID payload.

- The certificates in the certificate payload shall be encoded as type 4 (X.509 Certificate – Signature).

7.1.5 TLS Usage Profile

TLS is used between FAP and the FMS in case when the FMS is in the public Internet or when end-to-end security is desired even in case when the FMS is located within the operator's intranet and there is IPsec tunnel between the FAP and the security gateway. In the end-to-end case, the TLS will be encapsulated inside the IPsec tunnel between the FAP and security gateway and will be normal TLS between the security gateway and the FMS within the operator's network. In addition to the forbidden use of shared-secret in TLS, the following requirements or restrictions are in place:

- The FMS acting as the TLS server shall always send the FMS TLS certificate in the ServerCertificate message

- The FAP acting as the TLS client shall send its FAP TLS certificate in the Certificate message if requested by the FMS

7.2 IKEv2 Example Call Flow Used in 3GPP

One common procedure in all of the Femtocells based on different access technology (3GPP, CDMA2000, and WiMAX) is the IKEv2-based certificate authentication. IKEv2 is defined in IETF RFC 5996. The certificate based mutual authentication is performed between the FAP and Security Gateway at the edge of the core network. IKEv2 is an Internet Key Exchange protocol designed to set up a security association in the IPsec. IKEv2 is an improvement of the original IKE protocol. In the context of Femtocells, it uses a certificate (either vendor certificate or operator issued certificate) for authentication and a Diffie-Hellman key exchange to establish the session keys (fir ciphering and integrity protection) used in IPsec. The following is an example Femto Authentication call flow based on the 3GPP Femto as specified in 3GPP's TS 33.320. Also included in the example is the optional EAP-AKA-based hosting party authentication exchange that is exclusive to the 3GPP Femtocells. Though the hosting party authentication exchange is not part of the core Femto authentication, but it shows the robustness of the IKEv2 protocol to be able to carry on multiple authentication exchanges within the same session. The call flows shows the AAA-server and the HSS that are used in the 3GPP network for AKA/EAP-AKA-based authentication. For clarity purposes, the term UE is also referred to as a mobile station or a MS in CDMA- and WiMAX-based technologies and H(e)NB in the 3GPP is the same as the FAP in CDMA Femtocells and WFAP in WiMAX Femtocells respectively. Additionally, and for completeness, the call flow also shows at the beginning a secure boot and device integrity process as it is mandatory to complete before the IKEv2 exchanges take place. As previously mentioned in Chapter 6, the co-located L-GW IP address may be assigned by the SeGW, which is also shown in the call flow below. Specifically, the L-GW IP address assignment is carried in the IKEv2 CFG payload.

1. As part of the initialization and power on process, the trusted environment brings H(e)NB to secure boot and performs device integrity check of H(e)NB where all the reference values of the modules to be loaded are cryptographically calculated and verified against the set of values that are securely stored in memory. If the device integrity check fails during the initialization and boot process, IKEv2 procedure is not performed and the process terminates.

2. Following successful device integrity check, the H(e)NB sends an IKE_SA_INIT request to the SeGW and this requests officially starts the IKE exchange process. Included in the IKE_SA_INIT request are nonces and Diffie-Hellman values and is used to negotiate security parameters for the subsequent IKE Security Association

3. The SeGW sends IKE_SA_INIT response, requesting a certificate from the H(e)NB. The SeGW indicates that it support Multiple Authentication by including the MULTIPLE_AUTH_SUPPORTED payload. The MULTIPLE_AUTH_SUPPORTED payload is necessary because the optional host party EAP-AKA authentication is also to be performed and this payload signals to the SeGW that before the current authentication procedure is finished, there is yet another authentication that needs to complete.

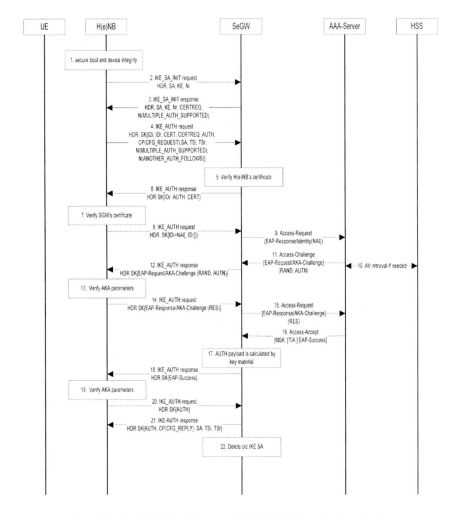

Figure 7.1. Combined certificate and EAP-AKA-based authentication

4. The H(e)NB inserts its identity in the IDi payload in this first
   message of the IKE_AUTH phase, computes the AUTH parameter
   within the secure execution environment of the trusted environment
   in the H(e)NB, and begins negotiation of child security associations.
   The authentication type indication in the user profile which is
   selected by H(e)NB's identity presented in the IDi payload may be
   used and enforce the choice of authentication (device only or
   combined device and HP). In other access technologies, this is not

necessary as only one type of authentication is supported and there is no such thing as hosting party authentication. The H(e)NB then sends IKE_AUTH request with the AUTH payload, its own certificate (either a vendor certificate or an operator certificate depending on whether the operator certificate has been enrolled into the H(e)NB or not), and also requests a certificate from the SeGW. Configuration payload is carried in this message if the H(e)NB's and/or L-GW's remote IP address(es) should be configured dynamically, as in the case the SeGW assigns a separate IP address for the L-GW. The H(e)NB indicates that it support Multiple Authentication (i.e. hosting party authentication) and that it wants to do a second authentication by including the MULTIPLE_AUTH_ SUPPORTED and ANOTHER_AUTH_FOLLOWS attributes. If configured to check the validity of the SeGW certificate the H(e)NB retrieves SeGW certificate status information from the OCSP responder. Alternatively the H(e)NB may add an OCSP request to the IKE message if online OCSP checking or cached response is needed explicitly for some reasons. Additionally, H(e)NB may also insert a N payload (not shown in the call flow) with a Notification type of INTEGRITY_INFO in the IKE_AUTH request for the purpose of carrying addition integrity related information to send to the SeGW.

5. The SeGW checks the correctness of the AUTH received from the H(e)NB and calculates the AUTH parameter which authenticates the second IKE_SA_INIT message. The authentication of the second IKE_SA_INIT message is a designed feature of the IKEv2 exchange. The SeGW verifies the certificate received from the H(e)NB. The SeGW may check the validity of the certificates using CRL or OCSP. If the H(e)NB request contained an OCSP request, or if the SeGW is configured to provide its certification revocation status to the H(e)NB, the SeGW retrieves SeGW certificate status information from the OCSP server, or uses a valid cached response if one is available. Online OCSP checking requires that the OCSP server be available to perform this checking and depending on the availability of the OCSP server, this may not always be possible. This OCSP or CRL check also serves as an device authorization for the H(e)NB.

6. The SeGW sends IKE_AUTH response with its identity in the IDr payload, the AUTH parameter and its certificate to the H(e)NB. If the SeGW has SeGW certificate status information available, this information is added to the IKE response to H(e)NB. Otherwise, the information for the SeGW certificate status is omitted. Additionally, if Step 4 includes a N payload with additional integrity related information, the SeGW processes is based on local policy of the operator. The SeGW may choose to retain the integrity related information carried in the N payload for statistical analysis, send the information to a Fraud Information Gathering System (FIGS) for fraud detection, or send the information to a remote validation entity for further validation.

7. The H(e)NB verifies the SeGW certificate with its stored root certificate. The root certificate for the SeGW certificate shall be stored in the secure memory within the trusted environment. The H(e)NB checks that the SeGW identity as contained in the SeGW certificate equals the SeGW identity as provided to H(e)NB by the initial configuration or by Femto Management Server. The H(e)NB checks the validity of the SeGW certificates using the OCSP response if configured to do so or if the OCSP response is present.

8. The H(e)NB sends another IKE_AUTH request message with the hosting party's identity in the IDi payload to inform the SeGW that the H(e)NB want to perform an EAP-based authentication. This time, the AUTH payload is no longer needed in the EAP-based authentication and is therefore omitted in the IKE_AUTH request that is sent the second time around.

9. The SeGW sends the Authentication Request message with an empty EAP Attribute Value Pair (AVP) to the 3GPP AAA Server, containing the identity received in IKE_AUTH request message received in step 8. Empty EAP AVP is used since the SeGW knows that a 3GPP specific AKA authentication is requested and no other type of authentication (e.g. CHAP, PAP, etc.) is supported by the HSS.

10. The AAA Server shall fetch the subscription data and authentication vectors from HSS/HLR based on the identity presented that corresponds to the hosting party and H(e)NB pair. There should be a tight-coupling or binding of the identities of the hosting party and

the H(e)NB as the credentials presented in the hosting party module (HPM), though similar to that of the credentials in a normal USIM, is not to be used or mis-used as a regular-subscription USIM in a normal handset for the purpose of originating or receiving calls. The subscription data only pertains to the use of the hosting module as a tool for the H(e)NB for Femto-based services.

11. The AAA Server initiates the authentication challenge between the AAA and the hosting party module within the H(e)NB. The hosting party module is not shown separately as another entity, but for the context of the IKEv2 authentication with optional hosting party authentication, it is understood that the hosting party module is inserted into the H(e)NB securely.

12. The SeGW sends IKE_AUTH response to H(e)NB. The EAP message received from the AAA Server (EAP-Request/AKA-Challenge) is included in order to start the EAP procedure over IKEv2. Since the IKEv2 procedure is far from complete and is considered suspended from the point of view of the H(e)NB, the SeGW simply acts as a pass through for any traffic that is sent to the H(e)NB or the Hosting Party Module in the H(e)NB.

13. The H(e)NB processes the EAP challenge message and uses the Hosting Party Module for verification of the AUTN and also for generating the RES response parameters. Optionally, processing of the whole EAP challenge message, including verification of the received MAC with the newly derived keying material may be performed entirely within the H(e)NB's Hosting Party Module.

14. The H(e)NB/HPM sends the IKE_AUTH request with the EAP-Response/AKA-Challenge to the SeGW. Again, the SeGW is simply acting as a pass through during the second authentication of the multiple authentications.

15. The SeGW forwards the EAP-Response/AKA-Challenge message to the 3GPP AAA Server for further verification.

16. When all checks and verifications are performed successfully, the AAA Server sends the Authentication Answer including an EAP success and the key material to the SeGW. This key material

includes the MSK that has been generated during the authentication process.

17. The EAP Success message is forwarded to the H(e)NB over an IKEv2 in IKE_AUTH response. The AUTH payload is calculated using the key material that was just generated in the previous steps.

18. 18 The H(e)NB takes its own copy of the MSK as input to generate the AUTH parameter to authenticate the first IKE_SA_INIT message. Though it may seem like a backward process to be authenticating the first IKE_SA_INIT message at this stage, this is a designed security feature of the IKEv2 process since only at this stage, the H(e)NB has enough information to provide the authentication that has been previously absent. Optionally the computation of the AUTH parameter is performed within the H(e)NB's HPM depending on operator policies and configurations.

19. IKE_AUTH request with the AUTH parameter is sent to the SeGW.

20. The MSK received in step 16 is used by the SeGW to generate the AUTH parameters in order to authenticate the IKE_SA_INIT phase messages.

21. The SeGW checks the correctness of the AUTH received from the H(e)NB. The SeGW should send the assigned Remote IP address in the configuration payload (CFG_REPLY), if the H(e)NB requested for H(e)NB's and/or L-GW's Remote IP address through the CFG_REQUEST. If the SeGW allocates different remote IP addresses to the L-GW and to the H(e)NB, then the SeGW can include necessary information in order to differentiate the IP addresses assigned to the H(e)NB and the L-GW. This is needed to avoid any mis-configuration as traffic from the network is routed based on the IP addresses of the H(e)NB or L-GW. If the CFG_REPLY is not used, other mechanism to inform which IP address is to be used for H(e)NB or L-GW may be possible and is implementation specific. Then the IKE_AUTH response with AUTH parameter is sent to the H(e)NB together with the configuration payload, security associations and the rest of the IKEv2 parameters and the IKEv2 negotiation terminates.

22. If the SeGW detects that an old IKE SA for that H(e)NB already exists, it will delete the IKE SA and send the H(e)NB an INFORMATIONAL exchange with a Delete payload in order to delete the old IKE SA in H(e)NB.

Note that Step 4 and Step 6 of the above example call flow mentioned the possible use of a Notify payload to carry additional integrity information. This further illustrates the flexibility and robustness of the IKEv2 protocol in that even though the use of the Notify payload is not the core part of the IKEv2, but instead of additional exchanges, it can be embedded as part of the main IKEv2 exchanges to conserve the number messages and rounds of message exchanges. The usefulness of the N payload to carry additional integrity information mentioned here is to support other validation methods or to support the use of other validation mechanisms not explicitly mention in the 3GPP specifications. In some implementations where remote validation is supported in addition to the autonomous validation, it may be useful to perform additional validation checks that are not part of the autonomous validation. Autonomous validation is sometimes viewed as a quick validation as it only performs the essential validation checks and may not cover all modules to be loaded. Furthermore, autonomous validation is self-contained and is not capable of providing additional integrity related information to the network. Some of the additional integrity related information may be useful for the network, for example for statistical purposes or for fraud management purposes. The potential use of the Notify payload to explicitly carry integrity data was proposed to the IP Security Maintenance and Extension (ipsecme) working group of the IETF in an Internet Draft in 2009 in the early stages of Femtocell security design. In the proposed extension to IKEv2, a number of data values are proposed that are viewed as useful and essential for the operation of the Femtocells as the autonomous validation process for the Femtocells may be considered either incomplete or inadequate in the views of some of the designers of the Femtocell security as there are limitation as the number of components that can be validated may be limited due to the relatively self-contained nature of the autonomous validation. Additionally, if the autonomous validation experiences any problems, it may be difficult for the operator or service center to find out exact nature of the problem the system may be experience. For example, it is not clear whether any of the following are evaluated as part of the autonomous validation:

- The Base Band System
- Radio Frequency System
- Clock System
- Board Support Packet (BSP)
- Data Center
- Operation & Maintenance (O&M)
- Transport Protocol Component
- Transmission Control Module
- Switching/Forwarding Module

If the IKEv2 protocols had supported and defined the use of the N payload to carry, for example integrity measurement parameters of some or all of the above mentioned subsystems or sub modules of the Femtocells, these sub modules could have been validated in the background as the IKEv2 process continues for the FAP as validation of these by a remote validation server is transparent to the FAP because the process would not take away any resources from the FAP. As mentioned already, the N payload may also be designed to carry other information about the FAP, for example state information that can be used for statistical analysis purposes, validation results that can be used to indicate failure of non-critical components, or even other information that can indicate attempt to compromise the FAP which can be used by fraud information management systems or law enforcement systems such as a FIGS. Nevertheless, these useful features may be considered for future releases. Since the use of the notify payload was not standardized, any use of it in the fore mentioned manner would need to rely upon specific vendor implementation using vendor specific extensions as defined in the standards.

Part III

From Femtocells to Small Cells

8

From Femtocells to Small Cells

8.1 Small Cells

Ever since the term "Femtocell" was coined and adopted as an industry-recognized name synonymous with a home base station in 2005 and subsequent launch of commercial Femto service in 2007 by the US CDMA carrier Sprint, it has grown in popularity and in size. The latest number shows that there are currently close to 100 million Femtocells in service serving over half a billion users worldwide. Forecast shows that the number of Femtocells can grow to up to 120 to 150 million by the end of 2014. The popularity of Femtocells is not only because of the fact that they have small form-factor, many of which have the size of an average Wi-Fi access point and can be combined (e.g. co-located) with Wi-Fi access point, home router, cable or DSL modem, but also because that they can be easily, quickly, and inexpensively deployed by operators to expand network coverage. Femto Forum, formed by leading vendors and operators has taken up the cause to promote Femtocells. To reflect the ever changing landscape of Femtocells and the entire wireless industry, the Femto Forum has made a change to Small Cell Forum. Not only will the Small Cell Forum continue to promote Femtocells, but it also expanded the coverage to other form of small cells. In many ways, Small Cell Forum has united the operators, vendors, and the regulators to come together to ensure the interoperability, long term viability and success of the small cells. Since its inception, the Femto Forum and later, the Small Cell Forum has worked relentlessly with many operators, vendors, suppliers, and system integrators to not only educate people, but also promote the entire small cell ecosystem. Though the forum itself is not a standards producing organization (SDO) like 3GPP or 3GPP2, but through the efforts of the forum and its members, a number of requirements, guidelines, and reports have been produced and published, including many application interface specifications, market status reports, etc. Many of these requirements have driven the adoption of related requirements in specifications published in other standards producing organizations such as the 3GPP or 3GPP2. The contribution of the Small Cell Forum to the entire Femtocell and small cell ecosystem has been invaluable.

While there are many unofficial definition of a small cell, but a small cell is generally considered to have the following features:

- it is smaller than the traditional macro base station
- its capacity is (significantly) less than that of a traditional macro base station
- it operates on licensed spectrum
- it is managed by the operator
- it connects to the operator's backhaul
- it may be deployed in customer premise, enterprise, or operator owned or leased location

Small cells can be a Femtocell, a Picocell, a Metrocell, a Microcell, Minicell or sometime even an Umbrella Cell. There are a variety of small cells associated with different with different generation of wireless such as 2G, 3G or 4G, for example 2G/3G Microcells or 4G Mini eNodeB.

These small cells differ greatly in size and coverage. While there is no consensus on the official definitions of these small cells, but from a coverage point of view, a Femtocell covers approximately 12 to 20 meters, a Picocell 200 meters, a Microcell 2 kilometres, and so forth. The biggest difference between Femtocell and rest of the small cell is that Femtocells are home or enterprise based base station. Because of this major difference, the design and deployment of Femtocells must take on additional security consideration when compared to the rest of other small cells. Another major difference is that the Femtocells are typically plug-and-play ready, meaning that the deployment of such requires very little operator or technical interaction. Once the Femtocell is deployed, the self-configuration and self-organization nature of these home base stations would take over with the help from the appropriate servers. Neither user or operator need to provide any manual interaction before the Femtocell can be up and running. As for the other small cells such as Picocell and Metrocell and in terms of security of these small cells, their design principle, deployment constraints, and operational practices are much closer to that of the macro base stations. They still require dedicated installation personnel to perform the necessary preparation before the small cell can be deployed. For example, RF engineering, site survey, interferences must be taken into account. Other than the smaller form factor and less capacity, these small cells (except Femtocell) functions exactly like it bigger counterpart, namely the macro base station.

Take the example of a Picocell, many manufacturers and operators considers that Picocell's intended purpose is to provide extended or provide additional coverage in many locations that may not require the capacity of an additional macro base station or if the operator is not able to justify putting a more expensive macro base station. A typical Picocell's coverage area is 200 meters or less where as a macro base station covers two kilometers or more. Some of the more traditional and typical deployment of the Picocell would be in in-building such as shopping malls, train stations, or hospitals. Since a typical Picocell (for example deployed in a 3GPP network) connects back to the operator's core network through dedicated links via a base station controller (BSC) or via a radio network controller (RNC). Like the traditional deployment of macro base stations, there is no need to deploy a security gateway unlike the deployment of Femtocells. Since from the BSC/RNC to the operator's core network is all considered as part of the operator's trusted domain, additional security afforded by the security gateway becomes unnecessary. Without the need for the security gateway makes the design of the Picocell much simpler than that of a Femtocell. Many of the security considerations of the Femtocell may not apply to the Picocell, and the designers and manufactures would simply concentrate on reducing the form factor (e.g. smaller component, less powerful antenna, processor, fewer memory, etc.). The inexpensive nature of the small cells (compared to the macro base stations) makes the small cells very attractive for the operators.

However, in some non-traditional and non-typical deployment scenarios where the traditional backhaul is not available, the consideration for the backhaul becomes one of the challenges of for these small cell deployments, especially when the deployment site is at hard-to-reach area. When traditional backhaul is not available, different backhaul solutions are considered and these solutions are influenced by many

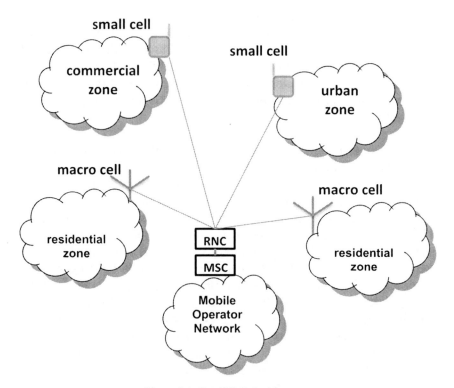

Figure 8.1. Small Cell Architecture

factors. Among some of these factors are operator's motivation to deploy small cells, capacity, cost effectiveness, size, availability, scalability, accessibility, and security. Some of the solutions that are available in the market include:

- Traditional microwave (6-42 GHz)
- Sub 6 GHz microwave – either point-to-point or point-to-multipoint
- "Light" licensed millimeter wave (E-band) microwave (70-80 GHz)
- Unlicensed millimeter wave (60 GHz)
- Copper wire xDSL
- Fiber optics
- Free space optics
- Cable / DOCSIS
- Satellite

As seen, it appears that there is an increasing trend of using high-speed wireless access technology (e.g. microwave) for backhaul. In the latest development, the US Federal Communications Commission (FCC) recently announced a change in its rules governing the 60 GHz (57– 64 GHz) band, making it one of the key technologies for LTE backhaul in an attempt to facilitate the use of this unlicensed spectrum as a backhaul alternative in densely-populated areas where 4G and other wireless services are experiencing an ever-increasing need for additional spectrum. Whatever the solution the operator chooses ultimately depends very much on the operator's overall deployment scenarios as well as the environment in which the small cells are deployed.

8.2 Small Cells and Wi-Fi

While on the subject of unlicensed spectrum, there had been some discussion of using Wi-Fi-based technology as the underlying access technology for Femtocells. However, recall that Wi-Fi operates on unlicensed spectrum, primarily over 2.4GHz ISM band and 5GHz bandwidths U-NII band, Wi-Fi access points using these frequencies would not be considered as Femtocells. One of the pre-requisite for being a Femtocell is that it has to operate on licensed spectrum. That aside, it does not prevent a Wi-Fi access point offer some of the similar services as the Femtocell, such as voice and Internet access. Another point worth mentioning is that the backbone of Wi-Fi is the Internet and the Internet service provider. Any similar services offered through the Wi-Fi are application based and there is little or no quality of service guarantee, such as for the voice services offered through a traditional wireless cellular service provider. Furthermore, there is no guarantee of interference-free operation among the different Wi-Fi access points as supplemented by many Internet service providers. As many 3G and 4G accesses are migrating away from the traditional circuit-based voice service, VOIP or Voice over IP has become the technology of choice for low cost voice services. However, when this voice service is offered through a wireless service provider, the quality of service guarantee still applies whereas the VOIP service offer service providers may or may not have the same level of service guarantee. Another example of such level of service guarantee is the mobility aspect. Wi-Fi typically does not support the full range of mobility (e.g. active handoff) that the users have become accustomed to with wireless services. In Wi-Fi, there is no seamless handoff as the users move from one access point to another access point. Many advanced have made transition from one access point to another less noticeable for the average users, especially for non-real-time Internet traffic, but the real-time voice, Wi-Fi still cannot offer the same seamless handover as the wireless network.

Nevertheless, Wi-Fi access is still invaluable in the proliferation of wireless broadband and has made communication anywhere and anywhere easier and more economical. It is a great supplement to the 3G and 4G services that are offered by the wireless service providers, not only in terms of better access and coverage, but it also made the user experience richer and more enjoyable as the users demand more and more bandwidth with faster and faster connection speed. To this extent, it is why carrier Wi-Fi is gaining traction as the operators increasingly use Wi-Fi as an ingenious tool to increase capacity and want to maintain direct control over the Wi-Fi infrastructure and at the same time share it with their roaming partners. The carrier Wi-Fi value proposition is especially attractive for the operators:

- Operators and partners own, operate, and manage the Wi-Fi network just as they have always done so with the wireless mobile network that they are operating today
- Advanced Wi-Fi functionalities to support easy access, faster authentication, increased security, and roaming support
- Near seamless switching back and forth between Wi-Fi coverage and existing wireless coverage
- Tightly integrated with existing framework and infrastructure such as standards established in 3GPP

As the industry alliances that are formed by leading operators and mobile equipment vendors work closely with the SDOs to ensure that the carrier Wi-Fi will not only integrate into the existing network infrastructure, but at the same time augment and expand the existing service offerings for the customers. One of such alliance is the Wi-Fi Alliance whose members include Huawei, Intel, TeliaSonera just to name a few. Rather than competing with the wireless industry, working closely with the entire wireless industry further cemented Wi-Fi's place for helping to drive the entire eco-system to realize the dream of achieving communication anytime and anywhere for users all over the world.

Part IV

Outlook and Concluding Remarks

9

Conclusion and Outlook

Indeed Femtocell has gotten off to a great start since the inception in the late 1990s when the idea of creating a dual mode home base station based on the GSM access technology and cordless telephony technology at the time. The idea was then that the home base station can be used to serve both cellular subscribers accessing via GSM terminals as well as home users accessing via a compatible cordless phone. While cost, among other reasons, almost made sure that the initial dual mode home base stations would not and could not succeed, great advances have been achieved during the next decade and the current decade. It made the next Femtocell a success story. Among the advances that made the Femto success story possible are mobile processing capabilities, Internet speed and capability breakthroughs, lower hardware costs, more compact hardware packaging, etc. In the pioneer days of Femtocells, some of the Femtocells were designed to work on low speed DSL or dial-up telephony Internet access with speed as low as 56 kilobits per second. That was especially true when some of the original business cases were made to provide low cost coverage extension and service expansion to rural areas or in third world countries where high-speed Internet was not readily available. Today, while the Internet speeds in rural areas and in third world countries are continuously improving, in more developed countries, it is more common to see much higher Internet speed in the 10's and 100's of megabits per second. This has made Femtocell services even better, more reliable, less costly to operate, and therefore more attractive.

The support of LIPA and SIPTO in Femtocell technology also made deploying Femto more attractive as the operators found more ways to offer additional services while users enjoy better quality of service and higher network speed at reasonably low costs. Offload is happening more and more when more Femtocells are deployed. However, as more and more LTE-based Femtocells are deployed, backhaul connection speed may play a role and negatively impact the service. Most LTE-based Femtocells are designed to support up to seven simultaneous users while the 3G Femtocells (e.g. UMTS) are only designed to support up to four simultaneous users. The consideration is then not to let the backhaul to become the bottleneck for the LTE Femtocells. As also with small cells, other access technologies are being used to lessen the impact of the backhaul.

According to the latest reports, there are already over 46 commercial services based on WiMAX, UMTS, CDMA, LTE access technologies covering over 3 billion mobile users around the world. That number continues to climb. Several major operators have estimated over 1 million units of Femtocells in service, though the numbers for other type of small cells are inevitably smaller as those are intended to target different deployment scenarios than the Femtocells. The forecast for both Femtocells and other small cells are looking very rosy indeed for the near future with numbers ranging anywhere from 50 million to 100 million of smalls shipped and/or deployed in the next two to three years. As more and more operators are expanding LTE coverage, LTE Femtocells have helped many operators to migrate from 3G to LTE much faster than the migration from 2G to 3G. Proliferation of smart phones and Wi-Fi-capable phones also drove the operators to deploy more small cells at a faster pace. Indeed in just a few short years, Femtocells and small cells have come a long way.

While the increase in Femtocells and Femtocells deployment makes for bigger targets for the hackers and attackers alike, security still remains on the minds of the users and operators. With the speed of information flow, social media, and communication anywhere and anytime, any small flaw or incident may spread quickly and get amplified out of proportion. Instead of rushing the Femtocells to market, great care has been placed in the design of the Femtocells and even greater care has been put into the security of the Femtocells. Operators, manufacturers, researchers, and users remain confident that the thoughts of hundreds and thousands of hours of countless individuals that were put in the design of the Femtocells to make sure that they are capable of withstand attackers and hackers are paying big dividend, even though the process may have taken longer than market had wanted. From time to time, there are still isolated reports of Femtocells being hacked, but those are extremely rare and chances are, those cases are most applicable to Femtocells that were manufactured prior to standardizations or due to improper implementations. While the open standards process can be both a friend and an enemy for the Femtocell. On one hand, all of the thoughts that have gone into the design are out in the open and on the other hand, it makes it both possible and easier for hackers and attackers to gain insight knowledge of the Femtocell. Rest assured, researchers, system designers, manufacturers, and operators are constantly improving the security of Femtocells, making the Femtocell success story a continued one for many years to come.

Annex

Contents provided in this section are to help the readers with some additional background information.

A.1 Sample of TR-069 FMS Security Related Parameters

TR-069Object/ Attribute Name	TR 069 Type	Permissible value range (if applicable)	Description
InternetGatewayDevice.DeviceInfo	Object		Standard TR069 Object
SoftwareVersion	string(64)	None	A string identifying the software version currently installed in the CPE.
HardwareVersion	Srting(64)	None	A string identifying the particular CPE model and version
SerialNumber	string(64)	None	A string identifying the particular CPE serial number.
Description	string(256)	None	Description of the CPE device (human readable string)
Model Name	string(64)	None	Model name of the CPE device (human readable string)
UpTime	UnsignedInt	None	Time in seconds since the CPE was last restarted.
First Use Date	dateTime	None	Date and time in UTC that the CPE first successfully established a network connection.

ProductClass	string(64)	None	Identifier of the class of product for which the serial number applies. That is, for a given manufacturer, this parameter is used to identify the product or class of product over which the SerialNumber parameter is unique.
Manufacturer	string(64)	None	The manufacturer of the CPE (human readable string).
ManufacturerOUI	string(6)	None	The manufacturer OUI of the device (human readable string)
ManufacturerOUI36	string(9)	None	36-bit Organizationally unique identifier of the device manufacturer as defined by EUI-64 (http://standards.ieee.org/regauth/oui/tutorials/EUI64.html). Represented as a nine hexadecimal-digit value using all upper-case letters and including any leading zeros. This value shall remain fixed over the lifetime of the device, including across firmware updates.
InternetGatewayDevice.ManagementServer	**Object**		**FSM configuration. Standard TR069 Object**
...			
Username	string(256)	none	Username used to authenticate the CPE when making a connection to the ACS using the CPE WAN Management

			Protocol.
Password	string(256)	none	Password used to authenticate the CPE when making a connection to the ACS using the CPE WAN Management Protocol.
...			
ConnectionRequestUserna me	string(256)	none	Username used to authenticate an ACS making a Connection Request to the CPE.
ConnectionRequestPasswor d	string(256)	none	Password used to authenticate an ACS making a Connection Request to the CPE. When read, this parameter returns an empty string, regardless of the actual value. When read, this parameter returns an empty string, regardless of the actual value.
.FAPservice.{i}.	**Object**		**This service object contains the 1X and HRPD Application service Objects. Standard TR069 Object**
DeviceType	String		The type of FAP device. Enumeration of: Standalone, Integrated
.FAPService.{i}.Capabiliti es	**Object**		**This service object contains the 1X and HRPD Application service Objects. Standard TR069 Object**
GPSEquipped	boolean	-	Indicates whether the FAP is equipped with a GPS receiver

			or not.
...			
MaxChildSAPerIKE	unsignedInt		Indicates the maximum number of child SAs per IKE session that the device is capable of supporting.
MaxIKESessions	unsignedInt		Indicates the maximum number of IKE sessions the device is capable of supporting at any given time.
...			
RemoteIPAccessCapable	boolean		Indicates whether the FAP is capable of providing remote IP access service as defined in 3GPP2 X.S0059
...			
EnclosureTamperingDetected	boolean		This parameter indicates whether or not physical tampering of the device enclosure occurred, such as illegal opening of the box. If *true* device tampering is detected. If *false* no sign of device tampering is detected. Tampering state must be persisted across reboots and the device MUST never reset it back from *true* to *false* even after a factory reset.
RemoteIPAccessEnable	boolean		Enable or disable remote IP access service as defined in 3GPP2 X.S0059.

...			
EmergencySessionPeriod	unsignedInt	15-1800	The period (in seconds) the session of an unauthorized user is kept alive after an emergency call.
MSCId	unsignedint	n/a	This is the FAP's MSCID. This ID is used in PANI header.
...			
SecGWServer1	string(64)		First SecGW the FAP attempts to contact with. Either FQDN or IP address
SecGWServer2	string(64)		Second SecGW the FAP attempts to contact with. Either FQDN or IP address
SecGWServer3	string(64)		Third SecGW that the FAP attempts to contact with. Can contain either FQDN or IP address
SecGWSwitchoverCount	UnsignedInt		Counter indicating how many times FAP has switched between SEGW
...			
AccessMode	unsignedInt	0- Open,1-Closed, 2-Signaling Association	Defines the access control mode of FAP. Default mode is Open, Allowed modes are Open, Closed and Signaling Association
...			
AccessControlList	string (1024)		Comma-separated list (maximum length 1024) of strings (maximum item length 15). Each entry is an IMSI.
...			

IMSI	String	Up to 15 characters	International Mobile Subscriber Identity of the UE.
MembershipExpires	dateTime		The time that the current ACL membership expires.
MaxNeighborListEntries	UnsignedInt	-	Maximum number of entries avaiable in .FAPService.{i}.Cell Config.CDMA.RAN. 1xRTT.NeighborList
NeighborListNumberofEntries	unsignedInt	-	The current number of entries in .FAPService.{i}.Cell Config.CDMA.RAN. 1xRTT.NeighborList
NghbrMaxAge	unsignedInt	0 - 15	neighbor set maximum age beyond which the mobiles are supposed to drop a neighbor. This is the count of neighbout list updat or extended neighbor list updates a mobile shall receive before removing a neighbor from the neighbor list. Reference-3GPP2 C.S0005-0 standards.
...			
BrdcastGpsAssit	UnsignedInt	0 - 1	1: Broadcast GPS assist is supported. Currently only a value of 0 is supported for this attribute. Reference-3GPP2 C.S0005-0 standards.
...			
OneXNeighborIndex	Int	(0-19)	Macro neighbor index. Reference-3GPP2 C.S0005-0 standards.
...			
baseIdentifier	Int	0	unique identifier of the neighboring cell

			(optional)
airInterface	string	IS-2000 IS95 Other	Air interface technology of the neighbor
...			
OneXNeighborLongitude	string	none	Longitude for this neighbor. Parameter is a string representing a floating point real number (+/- XXX.YY). Where + for East, - for West, XXX.Y ranges between 0.0 and 180.0
OneXNeighborLatitude	string	none	Latitude for this neighbor. Parameter is a string representing a floating point number (+/-XX.Y). + for North, - for South, XX.Y ranges between 0.0 and 90.0 .
OneXNeighborMSCId	unsignedInt	0-16777215	MSCID of the neighbors.
...			
batchTableSecurityParameterIndex	unsignedInt	256-4294967295	Security Parameter Index to be used
batchTableSecretKey	string	32 digit HEX	32 digit secret key
securityParameterIndex	unsignedInt	256..4294967295	Security Parameter Index used
securityKey	string	32 digit HEX	secret key
...			
registrationAttempts	unsignedInt32	none	Total number of registration attempts
registrationFails	unsignedInt32	none	Total number of registrations failed
registrationsBlocked	unsignedInt32	none	Unauthorized registrations blocked by CAC
pageAttempts	unsignedInt32	none	total number of attempts to page

			requests received from the core
pageFails	unsignedInt32	none	total number of page requests rejected or not responded
voiceCallAttempts	unsignedInt32	none	total number of voice call attempts
voiceCallFailures	unsignedInt32	none	total number of voice call attempts which failed for various reasons
voiceCallsBlocked	unsignedInt32	none	total number of voice calls blocked due to lack of authorization
voiceCallsDropped	unsignedInt32	none	Number of voice calls dropped due to various reasons other than a normal call release
dataCallAttempts	unsignedInt32	none	total number of data call attempts
dataCallFailures	unsignedInt32	none	total number of data call attempts failed for various reasons
dataCallsBlocked	unsignedInt32	none	total number of data calls failed due to lack of authorization
dataCallsDropped	unsignedInt32	none	total number of data calls dropped due to reasons other than a normal release.
averageVoiceCallInMsec	unsignedInt32	none	Average voice call duration in mili-seconds
averageDataCallInMsec	unsignedInt32	none	average data call duration
averageSessionInSec	unsignedInt32	none	Average session duration in seconds
totalVoiceCalls	unsignedInt32	none	Total number of successful voice calls
totalDataCalls	unsignedInt32	none	Total number of successful data calls
fwdVoicePacketDropPercentage	unsignedInt32	none	Percentage of voice packets dropped in MAC due to signaling
revVoicePacketDropPercentage	unsignedInt32	none	Percentage of voice packets received with

			bad FQI
fwdAvgDataRate	unsignedInt32	none	Average data rate of RB scheduler on FW link
revAvgDataRate	unsignedInt32	none	Average data rate of RN scheduler on reverse link
...			
batchTableSecurityParameterIndex	unsignedInt	256-4294967295	Security Parameter Index to be used
batchTableSecretKey	string	32 digit HEX	32 digit secret key
securityParameterIndex	unsignedInt	256..4294967295	Security Parameter Index used
securityKey	string	32 digit HEX	secret key
...			
anPppAuthenticationAttemptsByNode	unsignedInt	None	Number of AnPppAuthentications attempted by the Femtocell
a12RequestsSentFromNode	unsignedInt	None	Number of A12 Access Requests Sent from the Femtocell
a12RejectsReceivedByNode	unsignedInt	None	Number of A12 Access Rejects received from the Femtocell
a12AcceptsReceivedByNode	unsignedInt	None	Number of A12 Access Accepts received by the Femtocell
a12AccessChallengesReceivedByNode	unsignedInt	None	Number of A12 Access Challenges received by the Femtocell
a12RetransmitsSentFromNode	unsignedInt	None	Number of A12 Access Requests retransmitted from the Femtocell
...			
Impu	string (32)	none	IMPU value.
Impi	string (32)	none	Default IMPI value. Ignored in TNA mode.
SipPassword	string (32)	none	Password to Authenticate the IMS

			core; Ignored in TNA mode
SIPAuthCapability	string		comma-separated list of: TNA, SIPDigest
SIPAuthMethod	string	"TNA" or "SIPDigest"	The authentication method to be used by the FAP for SIP registration. TNA - use TNA for SIP authentication; SIPDigest (or empty string) - use SIP Digest for authentication, if SIPDigest is set, SIPDigestPassword must be set. In TNA mode, SIPPassword and Impi are ignored
SipUDPPort	unsignedInt	none	UDP port number for initiating SIP traffic.
RtpPortHighLimit	unsignedInt	none	The high limit of port number for originating and receiving RTP/RTCP traffic.
RtpPortLowLimit	unsignedInt	none	The low limit of port number for originating and receiving RTP/RTCP traffic.
SipRegExpiry	unsignedInt	none	SIP registration expiry value.
...			
.FAPService.{i}.Transport .Security	**object**		security object from TR-196
PkeyNumberOfEntries	unsignedInt		number of entries for Pkey object
CryptoProfileNumberOfEntries	unsignedInt		number of entries for CryptoProfile object
.FAPService.{i}.Transport .Security.Pkey.{i}	**object**		**Public key table of the FAP. Including all public-key certificates for the FAP.**

Enable	Boolean		whether enable this entry or not
LastModif	DateTime		The last time this entry was modified.
SerialNumber	string		The Serial Number field in an X.509 certificate
Issuer	string		The Issuer field in an X.509 certificate; i.e. the Distinguished Name (DN) of the entity who has signed the certificate.
NotBefore	DateTime		The beginning of the certificate validity period; i.e. the Not Before field in an X.509 certificate.
NotAfter	dateTime		the end of validity period
Subject	string		The X.501 DN of the entity associated with the public key.
SubjectAlt	string		The Subject AltName field in X.509 cert.
SignatureAlgorithm	string		The algorithm used in signing the certificate.
.FAPService.{i}.Transport .Security.CryptoProfile.{i }.	**Object**		This object contains parameters relating to IKEv2 and IPsec crypto profiles, which are essentially a subset of the typical IPsec SPD. RFC4301 [39]. At most one enabled entry in this table exists with all the same values for AuthMethod, IKEEncrypt, IKEPRF, IKEIntegrity, IKEDH, ESPEncrypt and ESPIntegrity.
Enable	boolean	FALSE	Enables and disables

			this entry.
AuthMethod	string	empty	Specifies the Security mechanism and set of credentials used by the FAP to authenticate itself. The value MUST be the full path name of a row in the .Transport.Security.P key table. If the referenced object is deleted, the parameter value MUST be set to an empty string. If an empty string, the FAP chooses the authentication method based on local policy. In order to configure the FAP for both FAP and hosting-party authentication, the object is populated with an enabled instance of the Pkey object.
MaxChildSA	unsignedInt [2, 4, 6, 8, 10]	2	Controls the maximum number of child SAs that can be negotiated by a single IKE session.
IKEEncrypt	string	AES-CBC	Comma-separated list of strings. IKEv2 encryption algorithm. Each list item is an enumeration of: 3DES-CBC AES-CBC

IKEIntegrtity	string	"HMAC-SHA1-96"	Comma-separated list of strings. IKEv2 integrity function. Each list item is an enumeration of: HMAC-SHA1-96 AES-XCBC-MAC-96
IKEPRF	string	HMAC-SHA1	Comma-separated list of strings. IKEv2 pseudorandom function. Each list item is an enumeration of: HMAC-SHA1 AES-XCBC-PRF-128
IKEDH	string	2048	Comma-separated list of strings. IKEv2 pseudorandom functionEach list item is an enumeration of: 1024 2048
ESPEncrypt	string	AESCBC	Comma-separated list of strings. IPsec encryption algorithm. Each list item is an enumeration of: 3DES-CBC AES-CBC Null
ESPIntegrity	string	HMAC-SHA1-96	Comma-separated list of strings. IPsec integrity function. Each list item is an enumeration of: HMAC-SHA1-96 AES-XCBC-MAC-96
IKERekeyLifetime	unsignedInt		IKE SA timeout in Seconds
IPSecRekeyLifetimeByte	unsignedint		IPSec SA timeout in kilobytes

IPSecRekeyLifetimeTime	unsignedInt		IPSec SA timeout in Seconds
DPDTimer	unsignedInt	300	DPD timeout in seconds.
NATKeepAliveTimer	unsignedInt	180	NAT-T keepalive timeout in seconds.
IPSecWindowSize	unsignedint	0	The size of the Anti-Replay Window. If 0 Sequence Number Verification is disabled.
SecMaxFragSize	unsignedint	256 - 1400 bytes	Maximum fragement size from FAP before Encryption. Sypported range is 256-1400. Default is 1200.
SecDFBit	unsignedint	0-Off, 1-on	Flag to indicate if the "do not fragment bit" needs to be turned on/off.
.FAPService.{i}.Transport .Tunnel.	**object**		
IKESANumberOfEntries	unsignedInt		The number of entries in the .Transport.Tunnel.IK ESA.{i}. table.
ChildSANumberOfEntries	unsignedInt		The number of entries in the .Transport.Tunnel.IK ESA.{i}. table.
VirtualInterfaceNumberOf Entries	unsignedInt		The number of entries in the .Transport.Tunnel.Vi rtualInterface.{i}. table.
MaxVirtualInterfaces	unsignedInt		The maximum number of virtual interfaces.
.FAPService.{i}.Transport .Tunnel.IKESA.{i}.	**object**		**IKE IPsec Security Association Table. This Table is a member of the IPsec Security Association Database (SAD). At most one entry in this table can exist with the same**

			values for IPAddress and SubnetMask.
Status	string		Enumeration:Disabled,Active,Completed, Progressing,Error
PeerAddress	string		IP address of the peer
CreationTime	dateTime		The time that the current IKE SA was set up.
IPAddress	string		the current IP address assigned by IKEv2 to this interface
SubnetMask	string		The current subnet mask assigned to this interface by IKEv2.
DNSServers	string		Comma-separated list (maximum length 256) of IPAddresses. Each item is an IP address of a DNS server for this interface assigned by IKEv2.
DHCPServers	string		Comma-separated list (maximum length 256) of IPAddresses. Each item is an IP address of a DHCP server for this interface assigned by IKEv2. A non empty list instructs the CPE to send any internal DHCP request to the address contained within this parameter.
IntegrityErrors	unsignedInt		The number of inbound packets discarded by the IKE SA due to Integrity checking errors.
OtherErrors	unsignedInt		The number of inbound packets discarded by the IKE SA due to other

			errors, such as anti-replay error.
AuthErrors	unsignedInt		The number of inbound packets discarded by the IKE SA due to authentication errors.
.FAPService.{i}.Transport .Tunnel.ChildSA.{i}.	**object**		**Child IPsec Security Association Table. This Table is a member of the IPsec Security Association Database (SAD). At most one entry in this table can exist with a given value for SPI.**
ParentID	unsignedInt		The value MUST be the instance number of a row in the .Transport.Tunnel.IK ESA table, or else be 0 if no row is currently referenced. If the referenced row is deleted, the parameter value MUST be set to 0.
SPI	unsignedInt		SPI value of the Child SA.
DirectionOutbound	Boolean		Traffic Direction. If true this Child SA refers to outbound traffic. If false this Child SA refers to inbound traffic.
CreationTime	DateTime		The time that the current Child SA was set up.
Traffic	unsignedInt		The measured traffic in bytes transferred by this ChildSA.
IntegrityErrors	unsignedInt		The number of inbound packets discarded by the Child SA due to Integrity

			checking errors.
ReplayErrors	unsignedInt		The number of inbound packets discarded by the Child SA due to anti-replay errors.
.FAPService.{i}.Transport .Tunnel.VirtualInterface.{ i}.	**object**		
Enable	Boolean		Enable or disable this entry.
CryptoProfile	string		The value MUST be the full path name of a row in the .Transport.Security.C ryptoProfile table. If the referenced object is deleted, the parameter value MUST be set to an empty string. If multiple instances of VirtualInterface point to the same CryptoProfile instance, the associated .Transport.Security.C ryptoProfile.{i}.Max ChildSA determines whether a new IKE session will be created (dynamically) to negotiate the child SA(s) for each of the virtual interfaces; otherwise, they are negotiated through the same IKE session.

DSCPMarkPolicy	int[i2:]		DSCP to mark the outer IP header for traffic that is associated with this virtual interface. A value of -1 indicates copy from the incoming packet. A value of -2 indicates automatic marking of DSCP as defined for the UMTS QoS class De-tunneled packets are never re-marked.
		-1	
...			
aGPSUserPW	string(64)	none	User password to be used with A-GPS request if needed.
...			
.FAPService.{i}.LoggingConfig	**Object**		**Logging configuration**
LogUploadServer	string(128)	none	FTP server host name or IP address which should be reachable from the Femto tunnel IP address to be used for log/statistics upload. If this is a null string, the Femto will not do any upload.
LogUploadServerPort	string(16)	none	Non-standard port range if needed.
LogUploadUserID	string(64)	none	User ID to be used with upload request
LogUploadUserPW	string(64)	none	User password to be used with upload request
DebugLogUploadPeriod	Int	0,1	0 disable periodic upload, 1 to enable periodic upload. LogUploadPeriod should be used to configure periodic upload. In release 2.2 this attribute should

			be used to periodically upload debug log. Supported ranges will be 300-2592000 (30 days). Default will be 1800. This default will take effect only after the log server details are configured. The unit will be seconds (0 means no upload).
DebugLogUploadNowReq	String	0,1	Set to 1 to trigger an immediate upload. Parameter is cleared when upload is completed. Supported values 0,1. Default value is 0.
FaultLogUploadPeriod	Int	0,1	0 disable periodic upload, 1 to enable periodic upload. LogUploadPeriod should be used to configure periodic upload. In release 2.2 this attribute should be used to periodically upload fault log. Supported ranges will be 300-2592000 (30 days). Default will be 3600. This default will take effect only after the log server details are configured. The unit will be seconds (0 means no upload).
FaultLogUploadNowReq	String	0,1	Set to 1 to trigger an immediate upload. Parameter is cleared when upload is completed. Supported values 0,1. Default value is 0.

ConnFailLogUploadPeriod	Int	0,1	0 disable periodic upload, 1 to enable periodic upload. LogUploadPeriod should be used to configure periodic upload. In release 2.2 this attribute should be used to periodically upload ConnFail log. Supported ranges will be 300-2592000 (30 days). Default will be 3600. This default will take effect only after the log server details are configured. The unit will be seconds (0 means no upload).
ConnFailLogUploadNowReq	String	0,1	Set to 1 to trigger an immediate upload. Parameter is cleared when upload is completed. Supported values 0,1. Default value is 0.
HealthLogUploadPeriod	Int	0,1	0 disable periodic upload, 1 to enable periodic upload. LogUploadPeriod should be used to configure periodic upload. In release 2.2 this attribute should be used to periodically upload Health log. Supported ranges will be 300-2592000 (30 days). Default will be 86400. This default will take effect only after the log server details are configured. The unit will be seconds (0 means no upload).

HealthLogUploadNowReq	String	0,1	Set to 1 to trigger an immediate upload. Parameter is cleared when upload is completed. Supported values 0,1. Default value is 0.
DebugLogGeneration	Integer	0,1	This attributes can be used to disable, enable debug log message collection. Default is enabled (1). Applicable values are 0-disabled, 1-enabled. If the periodic upload of the debug log is set and this attribute is disabled, the residual log gets uploaded in the next interval, and subsequent debug log uploads get disabled.
.FAPService.{i}.FaultMg mt.	**object**	**-**	**This object contains parameters relating to Fault/Alarm Management.**
SupportedAlarmNumberOf E ntries	unsignedInt	-	The number of entries in the *.FaultMgmt.Supporte dAlarm.{i}.* table.
MaxCurrentAlarmEntries	unsignedInt	-	The maximum number of entries allowed in the *.FaultMgmt.CurrentA larm.{i}.* table.
CurrentAlarmNumberOfEn tri es	unsignedInt	-	The number of entries in the *.FaultMgmt.CurrentA larm.{i}.* table.
HistoryEventNumberOfEnt ri es	unsignedInt	-	The number of entries in the *.FaultMgmt.HistoryE vent.{i}.* table.
ExpeditedEventNumberOf En tries	unsignedInt	-	The number of entries in the *.FaultMgmt.Expedite*

			dEvent.{i}. table.
QueuedEventNumberOfEntri es	unsignedInt	-	The number of entries in the *.FaultMgmt.QueuedEvent.{i}.* table.
.FAPService.{i}.FaultMgmt.CurrentAlarm.{i}.	**Object**		**Latest fault information**
AlarmIdentifier	string (64)	-	Identifies one Alarm Entry in the Alarm List. This value MUST be uniquely allocated by the FAP to the alarm instance during the lifetime of the individual alarm.
AlarmRaisedTime	dateTime	-	Indicates the date and time when the alarm was first raised by the FAP.
AlarmChangedTime	dateTime	-	Indicates the date and time when the alarm was last changed by the FAP.
ManagedObjectInstance	string (512)	-	Specifies the instance of the Informational Object Class in which the FAP alarm occurred by carrying the Distinguished Name (DN) of this object instance. This object may or may not be identical to the object instance actually emitting the notification to the ACS. The .DNPrefix should be pre-pended to the local DN to create the ManagedObjectInstance.

			Encode the Managed Objects representation
EventType	string(64)	-	Indicates the type of FAP event.
ProbableCause	string (64)	-	Qualifies the alarm and provides further information than EventType.
SpecificProblem	string (128)	-	Provides further qualification on the alarm beyond EventType and ProbableCause.This will be an empty string if the FAP doesn't support inclusion of this information.
PerceivedSeverity	string	-	Indicates the relative level of urgency for operator attention. Enumeration of: Critical Major Minor Warning Indeterminate (OPTIONAL) Although Indeterminate is defined in ITU-X.733. it SHOULD NOT be used by the FAP as a PerceivedSeverity.
AdditionalText	string (256)	-	This provides a textual string which is vendor defined. This will be an empty string if the FAP doesn't support inclusion of this information.
AdditionalInformation	string (256)	-	This contains additional information about the

			alarm and is vendor defined.

A.2 WiMAX SON

Self Organizing Networks by definition provide great automation in the management, operation, and configuration of Femtocells in WiMAX. The SON has the following functions in WiMAX:

- Self Configuration that provides initial radio related configurations
- Self Optimization that provides dynamic radio re-configuration, interference management, performance data collection, converage capacity optimization, mobility robustness optimization, and mobility load balancing
- Self Healing that provided fault detection localization and automatic fault correction

Self-configuration plays an important role in WFAP's initial configurations. Because of the some of the management functions of SON, sometimes, the functionality can be split across the FMS and SON or sometimes, they are co-located. Other self-configuration related functions include location verification, automatic configuration system identifier and parameters, automatic neighbor discovery, and auto-configuration of physical radio parameters. As part of the WFAP's self initialization and self-configuration process, WFAP must go through location discovery, SON system discovery, WFAP location authorization and self configuration, but not necessarily in the order listed above.

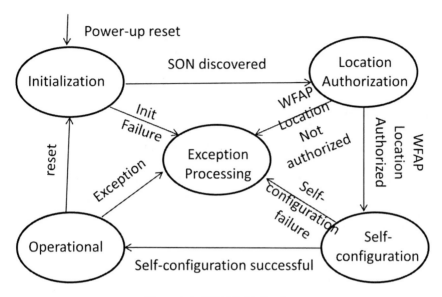

Figure A.1. WFAP Initialization

Self-optimization utilized real-time data gathered in operation so that interference, capability, and load conditions can be analyzed and re-configured to provide the best operating condition not only for the WFAPs in the system. Each WFAP actively collects data regarding air interface measurements such as transmit power of neighbors and signal to interference ratios. When a particular WFAP can no longer support a MS due to mobility or deterioration of air link conditions, mobility robustness optimization kicks in and tries to retain the MS, either through HO to a neighboring WFAP or a macro cell that has spare capacity or better air link condition. Mobility robustness optimization also tries to avoid late handovers by utilizing neighbor information collected. Some of the collected measurements include signal strength of WFAP neighbors, MS measurement reported, Femto HO parameters optimization and interference control, event based measurements such as cell specific call drops or HO failures, etc.

Self-healing mitigates faults automatically by monitoring alarms (Some alarm example related parameters can be found in the annex containing sample of TR-069 FMS Security Related Parameters) and alarm conditions. Regardless of whether faults are automatically cleared by self-healing, notifications are generated and logged so that appropriate personnel can review the events and determine any necessary course of action, for example notifying users that the WFAP cannot be brought to proper operating condition and therefore needs to be brought into service center for service and repair.

A.3 CDMA2000 Authentication

Prior to 3GPP2's adoption of the AKA protocol in CDMA2000 that was already used in 3G UMTS, earlier releases (Release A and B) of CDMA2000 needed to support backward compatibility of the older authentication method. This older authentication uses challenge response mechanism. There are two types of challenge-response mechanism used in CDMA: global challenge and unique challenge.

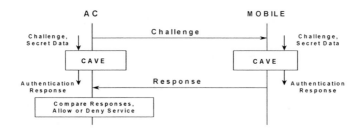

Figure A.2. CDMA2000 Challenge-Response Authentication

Essentially in a challenge-response mechanism, the system (AC in Figure A.2) sends out a challenge (in the form of a nonce or a random number), the mobile uses a key that it shares with the system (SSD or shared secret data) and computes an authentication response. The algorithm that is used in CDMA is the CAVE, or Cellular Authentication and Voice Encryption algorithm developed for 2G CDMA and continued to be used in early part of CDMA2000. The authentication response is sent back to the system and the system will compute the expected response and compare it to the one received from the mobile. If the two response match, the mobile is allowed access, otherwise, mobile access may be denied.

In addition to access, authentication may also be used in mobile registration, mobile termination, mobile data bursts, and TMSI assignment scenarios depending on operator policy.

A.3.1 Global Challenge

Global Challenge Authentication is a challenge-response mechanism, base on a globally broadcasted random number (RANDG) and is used to authenticate the MS on the access channels. When the global challenge authentication is active, the MS has to respond to the challenge (RANDG) with its authentication signature (AUTH) which is calculated based on the SSD-A, Mobile ID, the received challenge, and dialed digits (on call originations only). A successful response will permit the MS to access the cellular network (e.g., originate a call, respond to a page, or register in a new serving area). The broadcast frequency of the RANDG is controlled by the service provider based on the control channel bandwidth and the system processing capability. Global Challenge Authentication is also called the "main authentication mode."

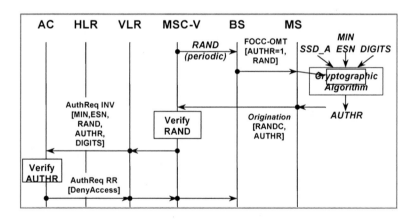

Figure A.3. CDMA2000 Global Challenge

A.3.2 Unique Challenge

Unique Challenge (UC) Response is a challenge-response type of mechanism based on a unique random number (RANDU) that is sent to a specific mobile rather than broadcast to all mobiles. It is essentially similar to the Global Challenge Response except that the response calculation is based not on a globally sent random number but on a unique random number sent whenever is needed. This procedure can be carried out either on the paging and access channels or on the forward and reverse traffic channels in CDMA. UC authentication can be used selectively by an operator to challenge a MS suspected of being compromised or is behaving badly, for testing purposes, or to authenticate an MS for accessing or requesting services in case a serving system cannot support Global Challenge Authentication for any reason.

Glossary

AAA	Authentication, Authorization and Accounting
AC	Authentication Center
ACL	Access Control lists
ACS	Access Control Server
AES	Advanced Encryption Standard
AH	Authentication Header
AHR	Access point Home Register
AK	Anonymity Key
AKA	Authentication and key agreement
AMF	Authentication Management Field
AMPS	Advanced Mobile Phone System
AN	Access Network
ARP	Address Resolution Protocol
AS	Access Stratum
ASME	Access Security Management Entity
ASN	Access Service Network
AT	Access Terminal
AuC	Authentication Center
AUTN	Authentication token
AV	Authentication Vector
BG	Border Gateway
BS	Base Station
BSP	Board Support Packet
CA	Certification Authority
CATT	Chinese Academy of Telecommunications Technology
CAVE	Cellular Authentication Voice Encryption
CC	Communication Content
CDMA	Code Division Multiple Access
Cell-ID	Cell Identity
CFG	Configuration (payload)
CK	Cipher Key

CKSN	Cipher Key Sequence Number
CPE	Customer Premise Equipment
C-RNTI	Cell RNTI
CRL	Certificate Revocation List
CS	Circuit Switched
CSCF	Call Session Control Function
CSG	Closed Subscriber Group
CSN	Connectivity Service Network
DECT	Digital Enhanced Cordless Telecommunications
DES	Data Encryption Standard
DHCP	Dynamic Host Configuration Protocol
DIFFSERV	Differentiated Service
DNS	Domain Name System
DOCSIS	Data Over Cable Service Interface Specification
DoI	Domain of Interpretation
(D)DoS	(Distributed) Denial of Service
DSL	Digital Subscriber Line
DSCP	Differentiated Services Code Point
EAP	Extensible Authentication Protocol
EARFCN-DL	E-UTRA Absolute Radio Frequency Channel Number-Down Link
ECM	EPS Connection Management
EDGE	Enhanced Data for GSM Evolution
EEA	EPS Encryption Algorithm
EIA	EPS Integrity Algorithm
eKSI	Key Set Identifier in E-UTRAN
EMM	EPS Mobility Management
eNB	Evolved Node-B
EPC	Evolved Packet Core
EPS	Evolved Packet System
EPS-AV	EPS authentication vector
ESN	Electronic Serial Number
ESP	Encapsulating Security Payload
E-UTRAN	Evolved UTRAN
FA	Foreign Agent
FCC	Federal Communication Commission
FCS	Femtocell Convergence Server

FEID	Femto Electronic Identifier
FIGS	Fraud Information Gathering System
FLCS	Femtocell Legacy Convergence Server
FQDN	Fully Qualified Domain Name
GC	Global Challenge
GERAN	GSM EDGE Radio Access Network
GLR	Gateway Location Register
GPON	Gigabit-capable Passive Optical Network
GRE	Generic Routing Encapsulation
GPRS	General Packet Radio Service
(A-)GPS	(Assisted-)Global Positioning System
GSM	Global System for Mobile Communications
GTP	GPRS Tunnelling Protocols
GUTI	Globally Unique Temporary Identity
HE	Home Environment
HeNB	Home eNodeB
HeNB-GW	Home eNodeB Gateway
HFN	Hyper Frame Number
HLR	Home Location Register
HNB	Home NodeB
HNB-GW	Hone NodeB Gateway
HO	Hand Over or Hand Off
HRPD	High Rate Packet Data
HSPA	High Speed Packet Access
HSS	Home Subscriber Server
iDEN	integrated Digital Enhanced Network
IEEE	Institute of Electrical and Electronic Engineers
IESG	Internet Engineering Steering Group
IETF	Internet Engineering Task Force
IGMP	Internet Group Management Protocol
IK	Integrity Key
IKE	Internet Key Exchange
IKEv2	Internet Key Exchange version 2
IMEI	International Mobile Station Equipment Identity
IMEISV	International Mobile Station Equipment Identity and Software Version number
IMS	IP Multimedia Subsystem

IMSI	International Mobile Subscriber Identity
IMT	International Mobile Telecommunications
IOS	Interoperability Specification
IP	Internet Protocol
IPsec	IP security
IRAT	Inter-Radio Access Technology
IRI	Intercept Related Information
ISAKMP	Internet Security Association Key Management Protocol
ISM	Industrial Scientific & Medical
ISR	Idle Mode Signalling Reduction
ITU	International Telecommunication Union
IV	Initialization Vector
KDF	Key Derivation Function
KSI	Key Set Identifier
LEMF	Law Enforcement Monitoring Function
LSB	Least Significant Bit
LSM	Limited Service Mode
LTE	Long Term Evolution
MAC	Message Authentication Code (also Medium Access Control)
MAC-I	Message Authentication Code for Integrity
MACT	Message Authentication Code T used in AES CMAC calculation
MAP	Mobile Application Part
MAPSEC	Mobile Application Part Security
ME	Mobile Equipment
MGW	Media Gateway
MIN	Mobile Identification Number
MME	Mobility Management Entity
MNO	Mobile Network Operator
MS	Mobile Station
MSC	Mobile Switching Center
MSC-V	Mobile Switching Center – Visited network
MSIN	Mobile Station Identification Number
NAP	Network Access Provider

NAS	Non Access Stratum
NAS-MAC	Message Authentication Code for NAS for Integrity
NAT	Network Address Translator
NCC	Next hop Chaining Counter
NDS	Network Domain Security
NDS/IP	NDS for IP based protocols
NE	Network Entity
NH	Next Hop
NIST	National Institute of Standards and Technology
NMT	Nordic Mobile Telephone
NPDB	Number Portability Database
NSP	Network Service Provider
NTP	Network Time Protocol
OCSP	Online Certificate Status Protocol
OFDM	Orthogonal Frequency Division Multiplexing
OS	Operating System
OTA	Over-The-Air
OTASP	Over-The-Air Service Provisioning
PCEF	Policy Control Enforcement Function
PCRF	Policy Control and Charging Rules Function
PCI	Physical Cell Identity
PDC	Personal Digital Cellular
PDCP	Packet Data Convergence Protocol
PKI	Public Key Infrastructure
PLMN	Public Land Mobile Network
PPPoE	Point-to-Point over Ethernet
PRNG	Pseudo Random Number Generator
PSK	Pre-shared Key
PS	Packet Switched
P-TMSI	Packet- Temporary Mobile Subscriber Identity
PTP	Precision Time Protocol
PVE	Platform Validation Entity
QoS	Quality of Service
RAND	RANDom number
RAU	Routing Area Update
RNTI	Radio Network Temporary Identifier

RRA	Radio Resource Agent
RRC	Radio Resource Control
RTMI	Radio Telefono Mobile Integrato
RTP	Real Time Protocol
SA	Security Association
SAD(B)	Security Association Database
SCTP	Stream Control Transmission Protocol
SEG	Security Gateway
SeGW	Security Gateway
SGSN	Serving GPRS Support Node
SIM	Subscriber Identity Module
SIP	Session Initiation Protocol
SMC	Security Mode Command
SN	Serving Network
SN id	Serving Network identity
SoC	System-on-Chip
SON	Self Organizing Network
SPD	Security Policy Database (sometimes also referred to as SPDB)
SPF	Signed Packet Format
SPI	Security Parameters Index
SQN	Sequence Number
SRB	Source Route Bridge
SRVCC	Single Radio Voice Call Continuity
SSH	Secure Shell
SSL	Secure Socket Layer
S-TMSI	S-Temporary Mobile Subscriber Identity
TACS	Total Access Communication System
TAI	Tracking Area Identity
TAU	Tracking Area Update
TCP	Transmission Control Protocol
TDMA	Time Division Multiple Access
TD-SCDMA	Time Division Synchronized Code Division Multiple Access
TE	Terminal Equipment
TLLI	Temporary Logical Link Identifier
TLS	Transport Layer Security
TMSI	Temporary Mobile Subscriber Identity

TrE	Trusted Environment
TISPAN	Telecoms & Internet converged Services & Protocols for Advanced Networks
TrGW	Transition Gateway
TTA	Telecommunications Technology Association
UC	Unique Challenge
UDP	User Datagram Protocol
UE	User Equipment
UEA	UMTS Encryption Algorithm
UIA	UMTS Integrity Algorithm
UICC	Universal Integrated Circuit Card
UMTS	Universal Mobile Telecommunication System
uPnP	Universal Plug-and-play
U-NII	Unlicensed National Information Infrastructure
UP	User Plane
USIM	Universal Subscriber Identity Module
UTC	Coordinated Universal Time
UTRAN	Universal Terrestrial Radio Access Network
VPN	Virtual Private Network
WAN	Wide Area Network
WBS	WiMAX Femto Bootstrap Server
WFMS	WiMAX Femto Management System
XRES	Expected Response
ZUC	ZU Chongzhi (LTE ciphering algorithm standardized for use in China deployments

Bibliography

[1] 3GPP TR 33.820: 3GPP Feasibility Study on the Security of Home Node B (HNB) / Home evolved Node B (HeNB)

[2] 3GPP TS 22.220: Service Requirements for Home Node B (HNB) and Home eNode B (HeNB)

[3] 3GPP TS 25.467: UTRAN architecture for 3G Home Node B (HNB); Stage 2

[4] 3GPP TS 33.210: Network Domain Security (NDS); IP Network Layer Security

[5] 3GPP TS 33.310: Network Domain Security (NDS); Authentication Framework (AF)

[6] 3GPP TS 33.320: Security of Home Node B (HNB) / Home evolved Node B (HeNB)

[7] 3GPP TS 33.401: 3GPP System Architecture Evolution (SAE); Security architecture

[8] 3GPP TS 35.222: 3GPP Confidentiality and Integrity Algorithms EEA3 & EIA3; Document 2: ZUC specification

[9] 3GPP TS 36.300: Evolved Universal Terrestrial Radio Access (E-UTRA) and Evolved Universal Terrestrial Radio Access Network (E-UTRAN); Overall description; Stage 2

[10] 3GPP2 A.S0024: Interoperability Specification (IOS) for Femtocell Access Points

[11] 3GPP2 C.S0016: Over-the-Air Service Provisioning of Mobile Stations in Spread Spectrum Standards

[12] 3GPP2 S.R0126: System Requirements for Femtocell Systems

[13] 3GPP2 S.R0135: Network Architecture for cdma2000 Femtocell Enabled Systems

[14] 3GPP2 S.R0139: Femtocell System Overview for cdma2000 Wireless Communication Systems

[15] 3GPP2 S.S0132: Femtocell Security Framework

[16] 3GPP2 X.S0059-000: cdma2000 Femtocell Network: Overview

[17] 3GPP2 X.S0059-100-0: Femto Network Specification

[18] 3GPP2 X.S0059-100-A: cdma2000 Femtocell Network: Packet Data Network Aspects

[19] 3GPP2 X.S0059-200: cdma2000 Femtocell Network: 1x and IMS Network Aspects

[20] 3GPP2 X.S0059-400: cdma2000 Femtocell Network: 1x Supplemental Service

Aspects

[21] 3GPP2 X.S0063: 3GPP2 Femtocell Configuration Parameters

[22] Blackhat 2011: Femtocells: A Poisonous Needle in the Operator's Hay Stack

[23] Blackhat 2013: I can hear you now: Traffic Interception and Remote Mobile Phone Cloning with a Compromised CDMA Femtocell

[24] Broadband Forum TR-069: CPE WAN Management Protocol

[25] ITU-R Report M.2134: Requirements related to technical performance for IMT-Advanced radio interface(s)

[26] ETSI ES 282 004: Telecommunications and Internet Converged Services and Protocols for Advanced Networking (TISPAN); NGN functional architecture; Network Attachment Sub-System (NASS)

[27] ETSI ES 283 035: Telecommunications and Internet Converged Services and Protocols for Advanced Networking (TISPAN); Network Attachment Sub-System (NASS); e2 interface based on the DIAMETER protocol

[28] The Hacker's Choice Blog: Vodafone Femtocell Hack

[29] Sprint User Guide Sprint Airave User Guide

[30] IEEE Std 802.16: IEEE Standard for Local and Metropolitan area networks – Part 16: Air Interface for Broadband Wireless Access System

[31] IEEE Std 1588: Standard for a Precision Clock Synchronization for Networked Measurement and Control Systems

[32] IETF Internet Draft: Integrity Data Exchanges in IKEv2

[33] IETF RFC 3161: SIP: Session Initiation Protocol

[34] IETF RFC 3261

[35] IETF RFC 4187: Extensible Authentication Protocol Method for 3rd Generation Authentication and Key Agreement (EAP-AKA)

[36] IETF RFC 4303: IP Encapsulating Security Payload (ESP)

[37] IETF RFC 4945

[38] IETF RFC 5683: Password-Authenticated Key (PAK) Diffie-Hellman Exchange

[39] IETF RFC 5905: Network Time Protocol (NTP) Version 4

[40] IETF RFC 5996: Internet Key Exchange Protocol Version 2 (IKEv2)

[41] ETF RFC 6712: Internet X.509 Public Key Infrastructure – HTTP Transfer for the Certificate Management Protocol (CMP)

[42] Infosec 2010: Hacking Femtocells: A femtostep to the holy grail

[43] NIST FIPS-197: Specification for the Advanced Encryption Standards (AES)

[44] WiMAX Forum T33-118-R016: Architecture, Detailed Protocols and Procedures, Network Architecture Femtocell Core Specification

[45] WiMAX Forum T33-119-R016: Architecture, Detailed Protocols and Procedures, Femtocell Management Specification

[46] WiMAX Forum T33-120-R016: Architecture, Detailed Protocols and Procedures, Self-Organizing Networks

[47] WiMAX Forum T31-123-R016: Requirements for WiMAX Femtocell Systems

About the Author

Marcus Wong received his Master of Arts Degree in Computer Science from Queens College of City University of New York (USA). He has over 20 years of experience in the wireless network security field with AT&T Bell Laboratories, AT&T Laboratories, Lucent Technologies, and Samsung's Advanced Institute of Technology. He holds Certification of Information System Security Professional (CISSP) from the prestigious International Information Systems Security Certification Consortium (ISC2).

Marcus has concentrated his research and work in many aspects of security in wireless communication systems, including 2G/3G/4G mobile networks, Personal Area Networks, and satellite communication systems. Marcus joined Huawei Technologies (USA) in 2007 and continued his focus on research and standardization in 3GPP, WiMAX Forum, IEEE, and IETF security areas. As an active contributor in the Wireless World Research Forum (WWRF), he has shared his security research on a variety of projects contributing toward whitepapers, book chapters, and speaking engagements.

In the past, Marcus has held elected official positions in both WWRF and 3GPP, serving as the vice-Chairman of WWRF Working Group 7 (Security and Trust working group) from 2007 to 2012 and as the vice-Chairman of 3GPP SA3 (Service & System Aspect, Security Group) from 2009 to 2011 respectively. He also served as guest editor in the IEEE Vehicular Technology magazine. He also has published a number of journal papers and whitepapers in leading publications, including *Journal of Cyber Security and Mobility*. In addition, he has numerous patents granted and/or pending.

Lightning Source UK Ltd.
Milton Keynes UK
UKOW04n0609210414

230294UK00007B/26/P